Principles of Physics

Principles of Physics

Bonnie Russo

Larsen & Keller
www.larsen-keller.com

Principles of Physics
Bonnie Russo
ISBN: 978-1-64172-686-3 (Hardback)

⊟ Larsen & Keller

Published by Larsen and Keller Education,
5 Penn Plaza,
19th Floor,
New York, NY 10001, USA

Cataloging-in-Publication Data

Principles of physics / Bonnie Russo.
    p. cm.
Includes bibliographical references and index.
ISBN 978-1-64172-686-3
1. Physics. 2. Dynamics. 3. Physical sciences. I. Russo, Bonnie.
QC21.3 .P75 2022
530--dc23

For more information regarding Larsen and Keller Education and its products, please visit the publisher's website www.larsen-keller.com

Table of Contents

Preface

The natural science which focuses on the study of matter, its motion and behavior through space and time is known as physics. It also studies the related entities of force and energy. Physics is a vast field that aims to understand how the universe behaves. It also intersects with other significant areas such as biophysics and quantum chemistry. The core theories of physics include quantum mechanics, special relativity, classical mechanics, thermodynamics and electromagnetism. Physics can be divided into classical and modern physics. Classical physics delves into the study of matter and energy at normal scales of observation. Whereas modern physics studies them under extreme conditions. Some of the prominent subfields of physics include nuclear and particle physics, condensed matter physics, optical physics, atomic physics, astrophysics and molecular physics. This textbook attempts to understand the multiple branches that fall under the discipline of physics and how such concepts have practical applications. Such selected concepts that redefine this field have been presented in this textbook. It aims to serve as a resource guide for students and contribute to the growth of the discipline.

A short introduction to every chapter is written below to provide an overview of the content of the book:

Chapter 1 - Physics is a branch of science that deals with the study of matter. It also delves into its motion and behaviour through space and time. It is also concerned with the energy and force related to matter. This is an introductory chapter which will introduce briefly all the significant aspects of physics; **Chapter 2** - Classical mechanics is the s tudy o f m otion of macroscopic objects such as stars, planets, galaxies and projectiles. Kinematics, dynamics and statics are the three main branches that fall under this domain. The topics elaborated in this chapter will help in gaining a better perspective about classical mechanics; **Chapter 3** - Gravity is a force by which all things with mass and energy including planets, starts, and galaxies are brought toward one and another. A potential energy that a physical object with mass has due to gravity in relation to another massive object is known as gravitational potential energy. This chapter discusses in detail the theories and methodologies related to gravity; **Chapter 4** - Thermodynamics is a domain of physics that is concerned with heat and temperature and their relation to radiation, energy, work and properties of matter. Statistical mechanics is a subset that deals with the study of any physical system that has a large number of degrees of freedom. The chapter closely examines the key concepts of thermodynamics and statistical mechanics to provide an extensive understanding of the subject; **Chapter 5** - Electromagnetic deals with the study of electromagnetic force, which is a type of physical interaction that occurs between electrically charged particles. It is concerned with the study of effects related to charge such as electric current, electric field, electric potentials, etc. This chapter discusses in detail the theories related to electromagnetic theory; **Chapter 6** - The branch of physics that studies the properties and behaviour of light is known as optics. It studies its interactions with matter and the construction of instrument that detects it. The chapter closely examines the key concepts of optics to provide an extensive understanding of the subject; **Chapter 7** - Special relativity is a physical theory that explains the relationship between space and time. It is based on two postulates i.e. in all inertial frames of reference, the laws of physics are identical and the speed of light in a vacuum is the same for all observers. All the diverse principles of special relativity have been carefully analysed in this chapter; **Chapter 8** - The theory of physics that describes nature at the smallest scales of energy levels of atoms and subatomic particles is known as quantum mechanics. It is a fundamental theory in physics that is applied in quantum computing, electronics, cryptography, quantum theory and macroscale quantum effects. The diverse applications of quantum mechanics in the current scenario have been thoroughly discussed in this chapter; **Chapter 9** - Atom is the smallest fundamental unit of ordinary matter with the properties of a chemical element. Atomic physics is the branch of physics that deals with the study of atoms as an isolated system of electrons and atomic nucleus. This chapter explains all the diverse aspects of atomic physics in order to develop a better understanding of the subject matter; **Chapter 10** - Nuclear physics is a domain of physics that is concerned with the study of atomic nuclei, their constituents and interactions. It is applied in various fields such as nuclear power, nuclear medicine, magnetic resonance imaging, nuclear weapons, industrial and agricultural isotopes, etc. The aim of this chapter is to provide an easy understanding of the varied facets of nuclear physics; **Chapter 11** - The branch of physics that studies the nature of the particles that constitute matter and radiation is known as particle physics. Gravitational force, weak force, electromagnetic force, strong forces are the four fundamental forces that fall under the domain of particle physics. The topics elaborated in this chapter will help in gaining a better perspective about particle physics.

Finally, I would like to thank my fellow scholars who gave constructive feedback and my family members who supported me at every step.

Bonnie Russo

The Basics of Physics

Physics is a branch of science that deals with the study of matter. It also delves into its motion and behaviour through space and time. It is also concerned with the energy and force related to matter. This is an introductory chapter which will introduce briefly all the significant as-pects of physics.

Physics is the science that deals with the structure of matter and the interactions between the fundamental constituents of the observable universe. In the broadest sense, physics (from the Greek physikos) is concerned with all aspects of nature on both the macroscopic and submicroscopic levels. Its scope of study encompasses not only the behaviour of objects under the action of given forces but also the nature and origin of gravitational, electromagnetic, and nuclear force fields. Its ultimate objective is the formulation of a few comprehensive principles that bring together and explain all such disparate phenomena.

Physics is the basic physical science. Until rather recent times physics and natural philosophy were used interchangeably for the science whose aim is the discovery and formulation of the fundamental laws of nature. As the modern sciences developed and became increasingly specialized, physics came to denote that part of physical science not included in astronomy, chemistry, geology, and engineering. Physics plays an important role in all the natural sciences, however, and all such fields have branches in which physical laws and measurements receive special emphasis, bearing such names as astrophysics, geophysics, biophysics, and even psychophysics. Physics can, at base, be defined as the science of matter, motion, and energy. Its laws are typically expressed with economy and precision in the language of mathematics.

Both experiment, the observation of phenomena under conditions that are controlled as precisely as possible, and theory, the formulation of a unified conceptual framework, play essential and complementary roles in the advancement of physics. Physical experiments result in measurements, which are compared with the outcome predicted by theory. A theory that reliably predicts the results of experiments to which it is applicable is said to embody a law of physics. However, a law is always subject to modification, replacement, or restriction to a more limited domain, if a later experiment makes it necessary.

The ultimate aim of physics is to find a unified set of laws governing matter, motion, and energy at small (microscopic) subatomic distances, at the human (macroscopic) scale of everyday life, and out to the largest distances (e.g., those on the extragalactic scale). This ambitious goal has been realized to a notable extent. Although a completely unified theory of physical phenomena has not yet been achieved (and possibly never will be), a remarkably small set of fundamental physical laws appears able to account for all known phenomena. The body of physics developed up to about the turn of the 20th century, known as classical physics, can largely account for the motions of macroscopic objects that move slowly with respect to the speed of light and for such phenomena as heat, sound, electricity, magnetism, and light. The modern developments of relativity and quantum mechanics modify these laws insofar as they apply to higher speeds, very massive objects, and to the tiny elementary constituents of matter, such as electrons, protons, and neutrons.

Scope of Physics

Mechanics

Mechanics is generally taken to mean the study of the motion of objects (or their lack of motion) under the action of given forces. Classical mechanics is sometimes considered a branch of applied mathematics. It consists of kinematics, the description of motion, and dynamics, the study of the action of forces in producing either motion or static equilibrium (the latter constituting the science of statics). The 20th-century subjects of quantum mechanics, crucial to treating the structure of matter, subatomic particles, superfluidity, superconductivity, neutron stars, and other major phenomena, and relativistic mechanics, important when speeds approach that of light, are forms of mechanics.

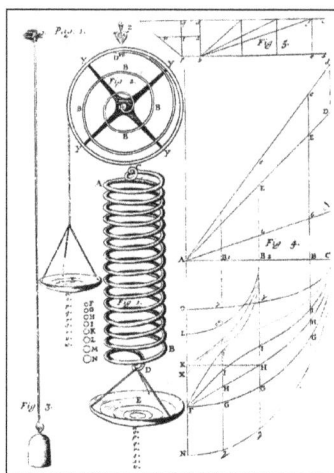

Illustration of Hooke's law of elasticity of materials, showing the stretching of a spring in proportion to the applied force, from Robert Hooke's *Lectures de Potentia Restitutiva* (1678).

In classical mechanics the laws are initially formulated for point particles in which the dimensions, shapes, and other intrinsic properties of bodies are ignored. Thus in the first approximation even objects as large as the Earth and the Sun are treated as pointlike—e.g., in calculating planetary orbital motion. In rigid-body dynamics, the extension of bodies and their mass distributions are considered as well, but they are imagined to be incapable of deformation. The mechanics of deformable solids is elasticity; hydrostatics and hydrodynamics treat, respectively, fluids at rest and in motion.

The three laws of motion set forth by Isaac Newton form the foundation of classical mechanics, together with the recognition that forces are directed quantities (vectors) and combine accordingly. The first law, also called the law of inertia, states that, unless acted upon by an external force, an object at rest remains at rest, or if in motion, it continues to move in a straight line with constant speed. Uniform motion therefore does not require a cause. Accordingly, mechanics concentrates not on motion as such but on the change in the state of motion of an object that results from the net force acting upon it. Newton's second law equates the net force on an object to the rate of change of its momentum, the latter being the product of the mass of a body and its velocity. Newton's third law, that of action and reaction, states that when two particles interact, the forces each exerts on the other are equal in magnitude and opposite in direction. Taken together, these mechanical laws in principle permit the determination of the future motions of a set of particles, providing their state of motion is known at some instant, as well as the forces that act between them and upon them from the outside. From this deterministic character of the laws of classical mechanics, profound (and probably incorrect) philosophical conclusions have been drawn in the past and even applied to human history.

Lying at the most basic level of physics, the laws of mechanics are characterized by certain symmetry properties, as exemplified in the aforementioned symmetry between action and reaction forces. Other symmetries, such as the invariance (i.e., unchanging form) of the laws under reflections and rotations carried out in space, reversal of time, or transformation to a different part of space or to a different epoch of time, are present both in classical mechanics and in relativistic mechanics, and with certain restrictions, also in quantum mechanics. The symmetry properties of the theory can be shown to have as mathematical consequences basic principles known as conservation laws, which assert the constancy in time of the values of certain physical quantities under prescribed conditions. The conserved quantities are the most important ones in physics; included among them are mass and energy (in relativity theory, mass and energy are equivalent and are conserved together), momentum, angular momentum, and electric charge.

Study of Gravitation

This field of inquiry has in the past been placed within classical mechanics for historical reasons, because both fields were brought to a high state of perfection by Newton and also because of its universal character. Newton's gravitational law states that every material particle in the universe attracts every other one with a force that acts along the line joining them and whose strength is directly proportional to the product of their masses and inversely proportional to the square of their separation. Newton's detailed accounting for the orbits of the planets and the Moon, as well as for such subtle gravitational effects as the tides and the precession of the equinoxes (a slow cyclical change in direction of the Earth's axis of rotation) through this fundamental force was the first triumph of classical mechanics. No further principles are required to understand the principal aspects of rocketry and space flight (although, of course, a formidable technology is needed to carry them out).

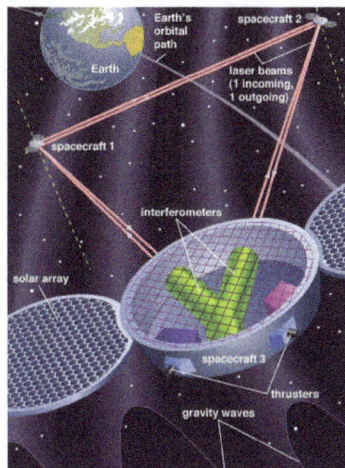

Laser Interferometer Space Antenna (LISA)

LISA, a Beyond Einstein Great Observatory, is scheduled for launch in 2015. Jointly funded by the National Aeronautics and Space Administration (NASA) and the European Space Agency (ESA), LISA will consist of three identical spacecraft that will trail the Earth in its orbit by about 50 million km (30 million miles). The spacecraft will contain thrusters for maneuvering them into an equilateral triangle, with sides of approximately 5 million km (3 million miles), such that the triangle's centre will be located along the Earth's orbit. By measuring the transmission of laser signals between the spacecraft (essentially a giant Michelson interferometer in space), scientists hope to detect and accurately measure gravity waves.

The modern theory of gravitation was formulated by Albert Einstein and is called the general theory of relativity. From the long-known equality of the quantity "mass" in Newton's second law of motion and that in his gravitational law, Einstein was struck by the fact that acceleration can locally annul a gravitational force (as occurs in the so-called weightlessness of astronauts in an Earth-orbiting spacecraft) and was led thereby to the concept of curved space-time. Completed in 1915, the theory was valued for many years mainly for its mathematical beauty and for correctly predicting a small number of phenomena, such as the gravitational bending of light around a massive object. Only in recent years, however, has it become a vital subject for both theoretical and experimental research.

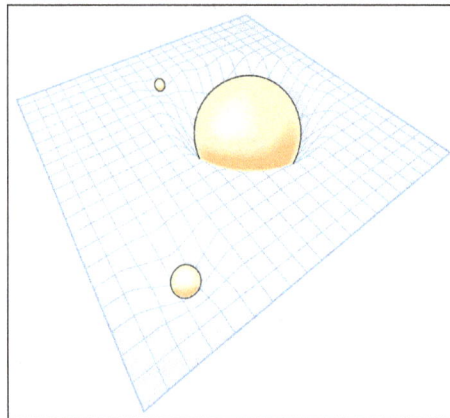

Curved space-time.

The four dimensional space-time continuum itself is distorted in the vicinity of any mass, with the amount of distortion depending on the mass and the distance from the mass. Thus, relativity accounts for Newton's inverse square law of gravity through geometry and thereby does away with the need for any mysterious "action at a distance."

The Study of Heat, Thermodynamics and Statistical Mechanics

Heat is a form of internal energy associated with the random motion of the molecular constituents of matter or with radiation. Temperature is an average of a part of the internal energy present in a body (it does not include the energy of molecular binding or of molecular rotation). The lowest possible energy state of a substance is defined as the absolute zero (−273.15 °C, or −459.67 °F) of temperature. An isolated body eventually reaches uniform temperature, a state known as thermal equilibrium, as do two or more bodies placed in contact. The formal study of states of matter at (or near) thermal equilibrium is called thermodynamics; it is capable of analyzing a large variety of thermal systems without considering their detailed microstructures.

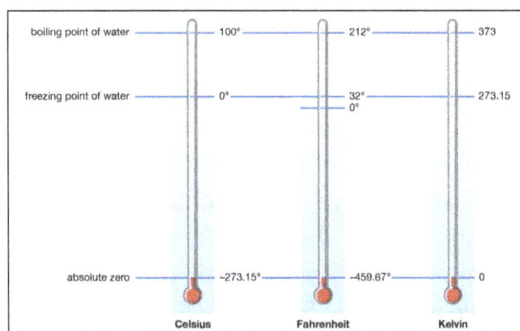

Standard and absolute temperature scales.

Statistical Mechanics

The science of statistical mechanics derives bulk properties of systems from the mechanical properties of their molecular constituents, assuming molecular chaos and applying the laws of probability. Regarding each possible configuration of the particles as equally likely, the chaotic state (the state of maximum entropy) is so enormously more likely than ordered states that an isolated system will evolve to it, as stated in the second law of thermodynamics. Such reasoning, placed in mathematically precise form, is typical of statistical mechanics, which is capable of deriving the laws of thermodynamics but goes beyond them in describing fluctuations (i.e., temporary departures) from the thermodynamic laws that describe only average behaviour. An example of a fluctuation phenomenon is the random motion of small particles suspended in a fluid, known as Brownian motion.

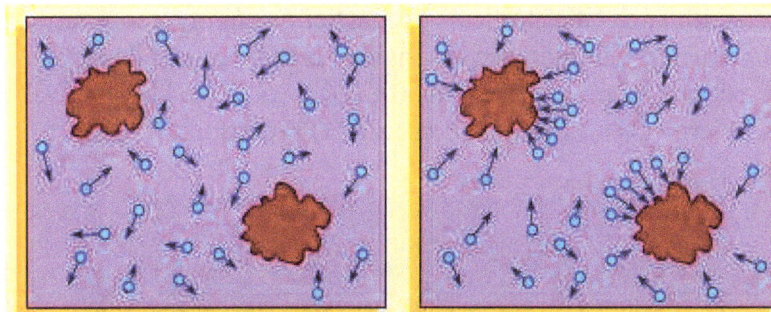

(Left) Random motion of a Brownian particle; (right) random discrepancy between the molecular pressures on different surfaces of the particle that cause motion.

Quantum statistical mechanics plays a major role in many other modern fields of science, as, for example, in plasma physics (the study of fully ionized gases), in solid-state physics, and in the study of stellar structure. From a microscopic point of view the laws of thermodynamics imply that, whereas the total quantity of energy of any isolated system is constant, what might be called the quality of this energy is degraded as the system moves inexorably, through the operation of the laws of chance, to states of increasing disorder until it finally reaches the state of maximum disorder (maximum entropy), in which all parts of the system are at the same temperature, and none of the state's energy may be usefully employed. When applied to the universe as a whole, considered as an isolated system, this ultimate chaotic condition has been called the "heat death."

Study of Electricity and Magnetism

Although conceived of as distinct phenomena until the 19th century, electricity and magnetism are now known to be components of the unified field of electromagnetism. Particles with electric charge interact by an electric force, while charged particles in motion produce and respond to magnetic forces as well. Many subatomic particles, including the electrically charged electron and proton and the electrically neutral neutron, behave like elementary magnets. On the other hand, in spite of systematic searches undertaken, no magnetic monopoles, which would be the magnetic analogues of electric charges, have ever been found.

The field concept plays a central role in the classical formulation of electromagnetism, as well as in many other areas of classical and contemporary physics. Einstein's gravitational field, for example, replaces Newton's concept of gravitational action at a distance. The field describing the electric force between a pair of charged particles works in the following manner: each particle creates an electric field in the space surrounding it, and so also at the position occupied by the other particle; each particle responds to the force exerted upon it by the electric field at its own position.

Classical electromagnetism is summarized by the laws of action of electric and magnetic fields upon electric charges and upon magnets and by four remarkable equations formulated in the latter part of the 19th century by the Scottish physicist James Clerk Maxwell. The latter equations describe the manner in which electric charges and currents produce electric and magnetic fields, as well as the manner in which changing magnetic fields produce electric fields, and vice versa. From these relations Maxwell inferred the existence of electromagnetic waves—associated electric and magnetic fields in space, detached from the charges that created them, traveling at the speed of light, and endowed with such "mechanical" properties as energy, momentum, and angular momentum. The light to which the human eye is sensitive is but one small segment of an electromagnetic spectrum that extends from long-wavelength radio waves to short-wavelength gamma rays and includes X-rays, microwaves, and infrared (or heat) radiation.

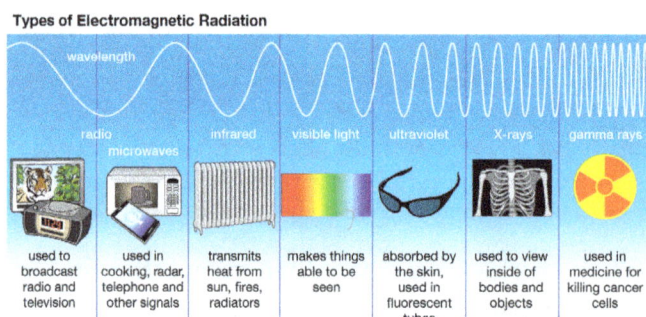

Radio waves, infrared rays, visible light, ultraviolet rays, X-rays, and gamma rays are all types of electromagnetic radiation. Radio waves have the longest wavelength, and gamma rays have the shortest wavelength.

Optics

Because light consists of electromagnetic waves, the propagation of light can be regarded as merely a branch of electromagnetism. However, it is usually dealt with as a separate subject called optics: the part that deals with the tracing of light rays is known as geometrical optics, while the part that treats the distinctive wave phenomena of light is called physical optics. More recently, there has developed a new and vital branch, quantum optics, which is concerned with the theory and application of the laser, a device that produces an intense coherent beam of unidirectional radiation useful for many applications.

Spectrum of white light by a diffraction grating. With a prism, the red end of the spectrum is more compressed than the violet end.

The formation of images by lenses, microscopes, telescopes, and other optical devices is described by ray optics, which assumes that the passage of light can be represented by straight lines, that is, rays. The subtler effects attributable to the wave property of visible light, however, require the explanations of physical optics. One basic wave effect is interference, whereby two waves present in a region of space combine at certain points to yield an enhanced resultant effect (e.g., the crests of the component waves adding together); at the other extreme, the two waves can annul each other, the crests of one wave filling in the troughs of the other. Another wave effect is diffraction, which causes light to spread into regions of the geometric shadow and causes the image produced by any optical device to be fuzzy to a degree dependent on the wavelength of the light. Optical instruments such as the interferometer and the diffraction grating can be used for measuring the wavelength of light precisely (about 500 micrometres) and for measuring distances to a small fraction of that length.

Atomic and Chemical Physics

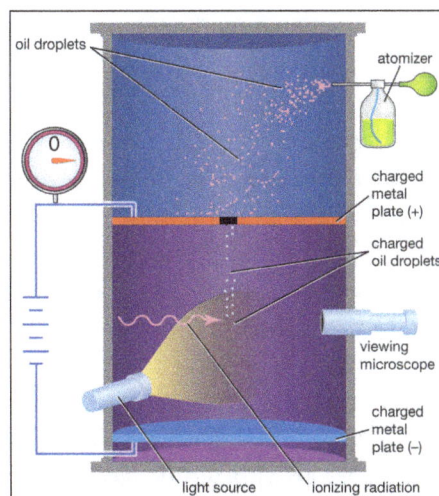

Millikan oil-drop experiment.

One of the great achievements of the 20th century was the establishment of the validity of the atomic hypothesis, first proposed in ancient times, that matter is made up of relatively few kinds of small, identical parts—namely, atoms. However, unlike the indivisible atom of Democritus and other ancients, the atom, as it is conceived today, can be separated into constituent electrons and nucleus. Atoms combine to form molecules, whose structure is studied by chemistry and physical chemistry; they also form other types of compounds, such as crystals, studied in the field of condensed-matter physics. Such disciplines study the most important attributes of matter (not excluding biologic matter) that are encountered in normal experience—namely, those that depend almost entirely on the outer parts of the electronic structure of atoms. Only the mass of the atomic nucleus and its charge, which is equal to the total charge of the electrons in the neutral atom, affect the chemical and physical properties of matter.

Between 1909 and 1910 the American physicist Robert Millikan conducted a series of oil-drop experiments. By comparing applied electric force with changes in the motion of the oil drops, he was able to determine the electric charge on each drop. He found that all of the drops had charges that were simple multiples of a single number, the fundamental charge of the electron.

Although there are some analogies between the solar system and the atom due to the fact that the strengths of gravitational and electrostatic forces both fall off as the inverse square of the distance, the classical forms of electromagnetism and mechanics fail when applied to tiny, rapidly moving atomic constituents. Atomic structure is comprehensible only on the basis of quantum mechanics, and its finer details require as well the use of quantum electrodynamics (QED).

Atomic properties are inferred mostly by the use of indirect experiments. Of greatest importance has been spectroscopy, which is concerned with the measurement and interpretation of the electromagnetic radiations either emitted or absorbed by materials. These radiations have a distinctive character, which quantum mechanics relates quantitatively to the structures that produce and absorb them. It is truly remarkable that these structures are in principle, and often in practice, amenable to precise calculation in terms of a few basic physical constants: the mass and charge of the electron, the speed of light, and Planck's constant (approximately $6.62606957 \times 10-34$ joule·second), the fundamental constant of the quantum theory named for the German physicist Max Planck.

Condensed-matter Physics

This field, which treats the thermal, elastic, electrical, magnetic, and optical properties of solid and liquid substances, grew at an explosive rate in the second half of the 20th century and scored numerous important scientific and technical achievements, including the transistor. Among solid materials, the greatest theoretical advances have been in the study of crystalline materials whose simple repetitive geometric arrays of atoms are multiple-particle systems that allow treatment by quantum mechanics. Because the atoms in a solid are coordinated with each other over large distances, the theory must go beyond that appropriate for atoms and molecules. Thus conductors, such as metals, contain some so-called free electrons, or valence electrons, which are responsible for the electrical and most of the thermal conductivity of the material and which belong collectively to the whole solid rather than to individual atoms. Semiconductors and insulators, either crystalline or amorphous, are other materials studied in this field of physics.

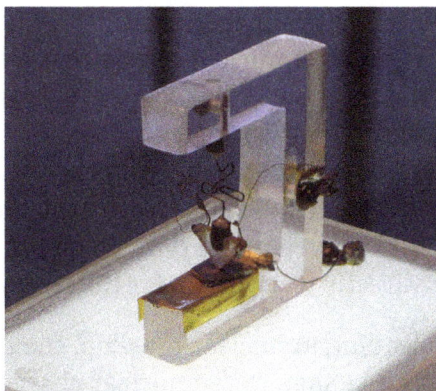

The first transistor, invented by American physicists John Bardeen,
Walter H. Brattain, and William B. Shockley.

Other aspects of condensed matter involve the properties of the ordinary liquid state, of liquid crystals, and, at temperatures near absolute zero, of the so-called quantum liquids. The latter exhibit a property known as superfluidity (completely frictionless flow), which is an example of macroscopic quantum phenomena. Such phenomena are also exemplified by superconductivity (completely resistance-less flow of electricity), a low-temperature property of certain metallic and ceramic materials. Besides their significance to technology, macroscopic liquid and solid quantum states are important in astrophysical theories of stellar structure in, for example, neutron stars.

Nuclear Physics

This branch of physics deals with the structure of the atomic nucleus and the radiation from unstable nuclei. About 10,000 times smaller than the atom, the constituent particles of the nucleus, protons and neutrons, attract one another so strongly by the nuclear forces that nuclear energies are approximately 1,000,000 times larger than typical atomic energies. Quantum theory is needed for understanding nuclear structure.

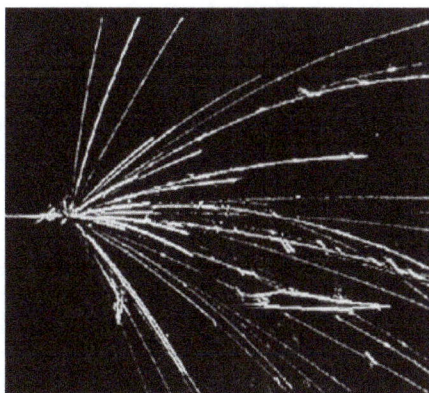

Particle tracks from the collision of an accelerated nucleus of a niobium atom with another niobium nucleus.
The single line on the left is the track of the incoming projectile nucleus, and the other
tracks are fragments from the collision.

Like excited atoms, unstable radioactive nuclei (either naturally occurring or artificially produced) can emit electromagnetic radiation. The energetic nuclear photons are called gamma rays. Radioactive nuclei also emit other particles: negative and positive electrons (beta rays), accompanied by neutrinos, and helium nuclei (alpha rays).

A principal research tool of nuclear physics involves the use of beams of particles (e.g., protons or electrons) directed as projectiles against nuclear targets. Recoiling particles and any resultant nuclear fragments are detected, and their directions and energies are analyzed to reveal details of nuclear structure and to learn more about the strong force. A much weaker nuclear force, the so-called weak interaction, is responsible for the emission of beta rays. Nuclear collision experiments use beams of higher-energy particles, including those of unstable particles called mesons produced by primary nuclear collisions in accelerators dubbed meson factories. Exchange of mesons between protons and neutrons is directly responsible for the strong force.

In radioactivity and in collisions leading to nuclear breakup, the chemical identity of the nuclear target is altered whenever there is a change in the nuclear charge. In fission and fusion nuclear reactions in which unstable nuclei are, respectively, split into smaller nuclei or amalgamated into larger ones, the energy release far exceeds that of any chemical reaction.

Particle Physics

One of the most significant branches of contemporary physics is the study of the fundamental sub-atomic constituents of matter, the elementary particles. This field, also called high-energy physics, emerged in the 1930s out of the developing experimental areas of nuclear and cosmic-ray physics. Initially investigators studied cosmic rays, the very-high-energy extraterrestrial radiations that fall upon the Earth and interact in the atmosphere. However, after World War II, scientists gradually began using high-energy particle accelerators to provide subatomic particles for study. Quantum field theory, a generalization of QED to other types of force fields, is essential for the analysis of high-energy physics. Subatomic particles cannot be visualized as tiny analogues of ordinary material objects such as billiard balls, for they have properties that appear contradictory from the classical viewpoint. That is to say, while they possess charge, spin, mass, magnetism, and other complex characteristics, they are nonetheless regarded as pointlike.

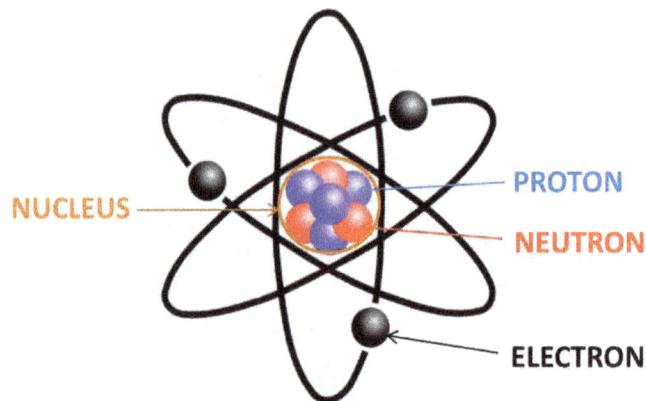

Very simplified illustrations of protons, neutrons, pions, and other hadrons show that they are made of quarks (yellow spheres) and antiquarks (green spheres), which are bound together by gluons (bent ribbons).

During the latter half of the 20th century, a coherent picture evolved of the underlying strata of matter involving two types of subatomic particles: fermions (baryons and leptons), which have odd half-integral angular momentum (spin $1/_2$, $3/_2$) and make up ordinary matter; and bosons(gluons, mesons, and photons), which have integral spins and mediate the fundamental forces of physics. Leptons (e.g., electrons, muons, taus), gluons, and photons are believed to be truly fundamental particles. Baryons (e.g., neutrons, protons) and mesons (e.g., pions, kaons), collectively known as hadrons, are believed to be formed from indivisible elements known as quarks, which have never been isolated.

Quarks come in six types, or "flavours," and have matching antiparticles, known as antiquarks. Quarks have charges that are either positive two-thirds or negative one-third of the electron's charge, while antiquarks have the opposite charges. Like quarks, each lepton has an antiparticle with properties that mirror those of its partner (the antiparticle of the negatively charged electron is the positive electron, or positron; that of the neutrino is the antineutrino). In addition to their electric and magnetic properties, quarks participate in both the strong force (which binds them together) and the weak force (which underlies certain forms of radioactivity), while leptons take part in only the weak force.

Baryons, such as neutrons and protons, are formed by combining three quarks—thus baryons have a charge of −1, 0, or 1. Mesons, which are the particles that mediate the strong force inside the atomic nucleus, are composed of one quark and one antiquark; all known mesons have a charge of −2, −1, 0, 1, or 2. Most of the possible quark combinations, or hadrons, have very short lifetimes, and many of them have never been seen, though additional ones have been observed with each new generation of more powerful particle accelerators.

The quantum fields through which quarks and leptons interact with each other and with themselves consist of particle-like objects called quanta (from which quantum mechanicsderives its name). The first known quanta were those of the electromagnetic field; they are also called photons because light consists of them. A modern unified theory of weak and electromagnetic interactions, known as the electroweak theory, proposes that the weak force involves the exchange of particles about 100 times as massive as protons. These massive quanta have been observed—namely, two charged particles, W^+ and W^-, and a neutral one, W^o.

In the theory of the strong force known as quantum chromodynamics (QCD), eight quanta, called gluons, bind quarks to form baryons and also bind quarks to antiquarks to form mesons, the force itself being dubbed the "colour force." (This unusual use of the term colour is a somewhat forced analogue of ordinary colour mixing.) Quarks are said to come in three colours—red, blue, and green. (The opposites of these imaginary colours, minus-red, minus-blue, and minus-green, are ascribed to antiquarks.) Only certain colour combinations, namely colour-neutral, or "white" (i.e., equal mixtures of the above colours cancel out one another, resulting in no net colour), are conjectured to exist in nature in an observable form. The gluons and quarks themselves, being coloured, are permanently confined (deeply bound within the particles of which they are a part), while the colour-neutral composites such as protons can be directly observed. One consequence of colour confinement is that the observable particles are either electrically neutral or have charges that are integral multiples of the charge of the electron. A number of specific predictions of QCD have been experimentally tested and found correct.

Quantum Mechanics

Although the various branches of physics differ in their experimental methods and theoretical approaches, certain general principles apply to all of them. The forefront of contemporary advances in physics lies in the submicroscopic regime, whether it be in atomic, nuclear, condensed-matter, plasma, or particle physics, or in quantum optics, or even in the study of stellar structure. All are based upon quantum theory (i.e., quantum mechanics and quantum field theory) and relativity, which together form the theoretical foundations of modern physics. Many physical quantities whose classical counterparts vary continuously over a range of possible values are in quantum theory constrained to have discontinuous, or discrete, values. Furthermore, the intrinsically deterministic character of values in classical physics is replaced in quantum theory by intrinsic uncertainty.

According to quantum theory, electromagnetic radiation does not always consist of continuous waves; instead it must be viewed under some circumstances as a collection of particle-like photons, the energy and momentum of each being directly proportional to its frequency (or inversely proportional to its wavelength, the photons still possessing some wavelike characteristics). Conversely, electrons and other objects that appear as particles in classical physics are endowed by quantum theory with wavelike properties as well, such a particle's quantum wavelength being inversely proportional to its momentum. In both instances, the proportionality constant is the characteristic quantum of action (action being defined as energy × time)—that is to say, Planck's constant divided by 2π, or $\hbar$.

In principle, all of atomic and molecular physics, including the structure of atoms and their dynamics, the periodic table of elements and their chemical behaviour, as well as the spectroscopic, electrical, and other physical properties of atoms, molecules, and condensed matter, can be accounted for by quantum mechanics. Roughly speaking, the electrons in the atom must fit around the nucleus as some sort of standing wave (as given by the Schrödinger equation) analogous to the waves on a plucked violin or guitar string. As the fit determines the wavelength of the quantum wave, it necessarily determines its energy state. Consequently, atomic systems are restricted to certain discrete, or quantized, energies. When an atom undergoes a discontinuous transition, or quantum jump, its energy changes abruptly by a sharply defined amount, and a photon of that energy is emitted when the energy of the atom decreases, or is absorbed in the opposite case.

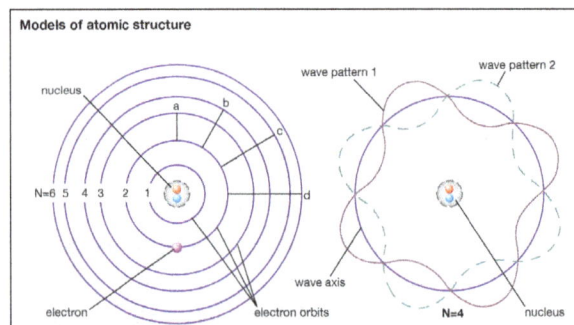

The Bohr theory sees an electron (left) as a point mass occupying certain energy levels. Wave mechanics sees an electron as a wave washing back and forth in the atom in certain patterns only. The wave patterns and energy levels correspond exactly.

Although atomic energies can be sharply defined, the positions of the electrons within the atom cannot be, quantum mechanics giving only the probability for the electrons to have certain locations. This is a consequence of the feature that distinguishes quantum theory from all other approaches to physics, the uncertainty principle of the German physicist Werner Heisenberg. This principle holds that measuring a particle's position with increasing precision necessarily increases the uncertainty as to the particle's momentum, and conversely. The ultimate degree of uncertainty is controlled by the magnitude of Planck's constant, which is so small as to have no apparent effects except in the world of microstructures. In the latter case, however, because both a particle's position and its velocity or momentum must be known precisely at some instant in order to predict its future history, quantum theory precludes such certain prediction and thus escapes determinism.

The complementary wave and particle aspects, or wave–particle duality, of electromagnetic radiation and of material particles furnish another illustration of the uncertainty principle. When an electron exhibits wavelike behaviour, as in the phenomenon of electron diffraction, this excludes its exhibiting particle-like behaviour in the same observation. Similarly, when electromagnetic radiation in the form of photons interacts with matter, as in the Compton effect in which X-ray photons collide with electrons, the result resembles a particle-like collision and the wave nature of electromagnetic radiation is precluded. The principle of complementarity, asserted by the Danish physicist Niels Bohr, who pioneered the theory of atomic structure, states that the physical world presents itself in the form of various complementary pictures, no one of which is by itself complete, all of these pictures being essential for our total understanding. Thus both wave and particle pictures are needed for understanding either the electron or the photon.

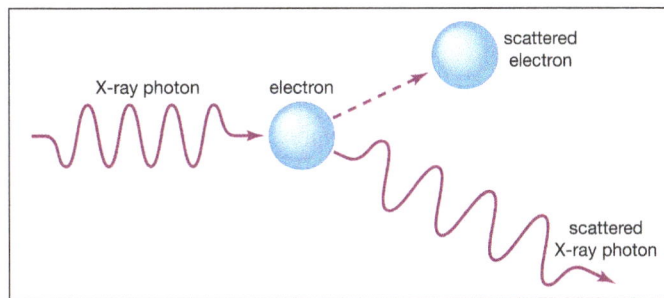

The Compton effect

When a beam of X-rays is aimed at a target material, some of the beam is deflected, and the scattered X-rays have a greater wavelength than the original beam. The physicist Arthur Holly Compton concluded that this phenomenon could only be explained if the X-rays were understood to be made up of discrete bundles or particles, now called photons, that lost some of their energy in the collisions with electrons in the target material and then scattered at lower energy.

Although it deals with probabilities and uncertainties, the quantum theory has been spectacularly successful in explaining otherwise inaccessible atomic phenomena and in thus far meeting every experimental test. Its predictions, especially those of QED, are the most precise and the best checked of any in physics; some of them have been tested and found accurate to better than one part per billion.

Relativistic Mechanics

In classical physics, space is conceived as having the absolute character of an empty stage in which events in nature unfold as time flows onward independently; events occurring simultaneously for one observer are presumed to be simultaneous for any other; mass is taken as impossible to create or destroy; and a particle given sufficient energy acquires a velocity that can increase without limit. The special theory of relativity, developed principally by Albert Einstein in 1905 and now so adequately confirmed by experiment as to have the status of physical law, shows that all these, as well as other apparently obvious assumptions, are false.

Specific and unusual relativistic effects flow directly from Einstein's two basic postulates, which are formulated in terms of so-called inertial reference frames. These are reference systems that move in such a way that in them Isaac Newton's first law, the law of inertia, is valid. The set of inertial frames consists of all those that move with constant velocity with respect to each other (accelerating frames therefore being excluded). Einstein's postulates are: (1) All observers, whatever their state of motion relative to a light source, measure the same speed for light; and (2) The laws of physics are the same in all inertial frames.

The first postulate, the constancy of the speed of light, is an experimental fact from which follow the distinctive relativistic phenomena of space contraction (or Lorentz-FitzGerald contraction), time dilation, and the relativity of simultaneity: as measured by an observer assumed to be at rest, an object in motion is contracted along the direction of its motion, and moving clocks run slow; two spatially separated events that are simultaneous for a stationary observer occur sequentially for a moving observer. As a consequence, space intervals in three-dimensional space are related to time intervals, thus forming so-called four-dimensional space-time.

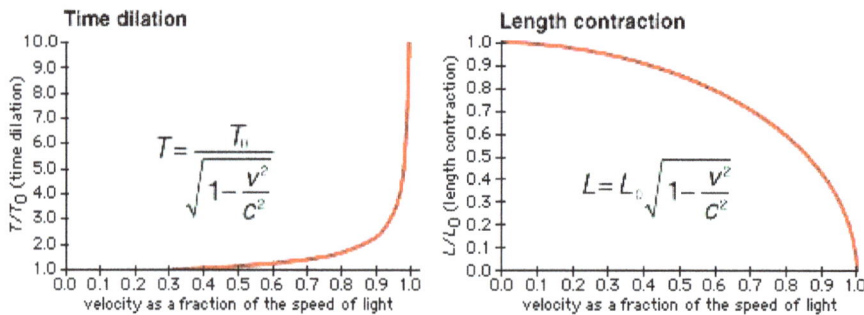

Length contraction and time dilation.

As an object approaches the speed of light, an observer sees the object become shorter and its time interval become longer, relative to the length and time interval when the object is at rest.

The second postulate is called the principle of relativity. It is equally valid in classical mechanics (but not in classical electrodynamics until Einstein reinterpreted it). This postulate implies, for example, that table tennis played on a train moving with constant velocity is just like table tennis played with the train at rest, the states of rest and motion being physically indistinguishable. In relativity theory, mechanical quantities such as momentum and energy have forms that are different from their classical counterparts but give the same values for speeds that are small compared to the speed of light, the maximum permissible speed in nature (about 300,000 kilometres per second, or 186,000 miles per second). According to relativity, mass and energy are equivalent and interchangeable quantities, the equivalence being expressed by Einstein's famous mass-energy equation $E = mc^2$, where m is an object's mass and c is the speed of light.

The general theory of relativity is Einstein's theory of gravitation, which uses the principle of the equivalence of gravitation and locally accelerating frames of reference. Einstein's theory has special mathematical beauty; it generalizes the "flat" space-time concept of special relativity to one of curvature. It forms the background of all modern cosmological theories. In contrast to some vulgarized popular notions of it, which confuse it with moral and other forms of relativism, Einstein's theory does not argue that "all is relative." On the contrary, it is largely a theory based upon those physical attributes that do not change, or, in the language of the theory, that are invariant.

Conservation Laws and Symmetry

Since the early period of modern physics, there have been conservation laws, which state that certain physical quantities, such as the total electric charge of an isolated system of bodies, do not change in the course of time. In the 20th century it has been proved mathematically that such laws follow from the symmetry properties of nature, as expressed in the laws of physics. The conservation of mass-energy of an isolated system, for example, follows from the assumption that the laws of physics may depend upon time intervals but not upon the specific time at which the laws are applied. The symmetries and the conservation laws that follow from them are regarded by modern physicists as being even more fundamental than the laws themselves, since they are able to limit the possible forms of laws that may be proposed in the future.

Conservation laws are valid in classical, relativistic, and quantum theory for mass-energy, momentum, angular momentum, and electric charge. (In nonrelativistic physics, mass and energy are separately conserved.) Momentum, a directed quantity equal to the mass of a body multiplied by its velocity or to the total mass of two or more bodies multiplied by the velocity of their centre of mass, is conserved when, and only when, no external force acts. Similarly angular momentum, which is related to spinning motions, is conserved in a system upon which no net turning force, called torque, acts. External forces and torques break the symmetry conditions from which the respective conservation laws follow.

In quantum theory, and especially in the theory of elementary particles, there are additional symmetries and conservation laws, some exact and others only approximately valid, which play no significant role in classical physics. Among these are the conservation of so-called quantum numbers related to left-right reflection symmetry of space (called parity) and to the reversal symmetry of motion (called time reversal). These quantum numbers are conserved in all processes other than the weak force.

Other symmetry properties not obviously related to space and time (and referred to as internal symmetries) characterize the different families of elementary particles and, by extension, their composites. Quarks, for example, have a property called baryon number, as do protons, neutrons, nuclei, and unstable quark composites. All of these except the quarks are known as baryons. A failure of baryon-number conservation would exhibit itself, for instance, by a proton decaying into lighter non-baryonic particles. Indeed, intensive search for such proton decay has been conducted, but so far it has been fruitless. Similar symmetries and conservation laws hold for an analogously defined lepton number, and they also appear, as does the law of baryon conservation, to hold absolutely.

Fundamental Forces and Fields

The four basic forces of nature, in order of increasing strength, are thought to be: (1) the gravitational force between particles with mass; (2) the electromagnetic force between particles with charge or magnetism or both; (3) the colour force, or strong force, between quarks; and (4) the weak force by which, for example, quarks can change their type, so that a neutron decays into a proton, an electron, and an antineutrino. The strong force that binds protons and neutrons into nuclei and is responsible for fission, fusion, and other nuclear reactions is in principle derived from the colour force. Nuclear physics is thus related to QCD as chemistry is to atomic physics.

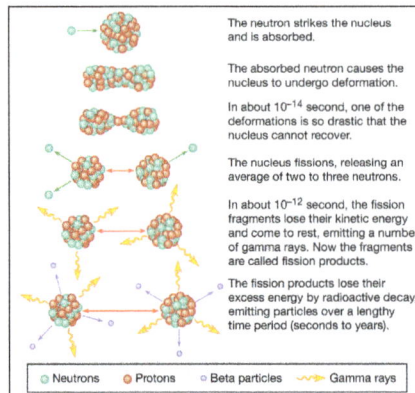

The neutron strikes the nucleus and is absorbed.

The absorbed neutron causes the nucleus to undergo deformation.

In about 10^{-14} second, one of the deformations is so drastic that the nucleus cannot recover.

The nucleus fissions, releasing an average of two to three neutrons.

In about 10^{-12} second, the fission fragments lose their kinetic energy and come to rest, emitting a number of gamma rays. Now the fragments are called fission products.

The fission products lose their excess energy by radioactive decay, emitting particles over a lengthy time period (seconds to years).

○ Neutrons   ● Protons   ○ Beta particles   ∿ Gamma rays

Sequence of events in the fission of a uranium nucleus by a neutron.

According to quantum field theory, each of the four fundamental interactions is mediated by the exchange of quanta, called vector gauge bosons, which share certain common characteristics. All have an intrinsic spin of one unit, measured in terms of Planck's constant ℏ. (Leptons and quarks each have one-half unit of spin.) Gauge theory studies the group of transformations, or Lie group, that leaves the basic physics of a quantum field invariant. Lie groups, which are named for the 19th-century Norwegian mathematician Sophus Lie, possess a special type of symmetry and continuity that made them first useful in the study of differential equations on smooth manifolds (an abstract mathematical space for modeling physical processes). This symmetry was first seen in the equations for electromagnetic potentials, quantities from which electromagnetic fields can be derived. It is possessed in pure form by the eight massless gluons of QCD, but in the electroweak theory—the unified theory of electromagnetic and weak force interactions—gauge symmetry is partially broken, so that only the photon remains massless, with the other gauge bosons (W^+, W^-, and Z) acquiring large masses. Theoretical physicists continue to seek a further unification of QCD with the electroweak theory and, more ambitiously still, to unify them with a quantum version of gravity in which the force would be transmitted by massless quanta of two units of spin called gravitons.

Methodology of Physics

Physics has evolved and continues to evolve without any single strategy. Essentially an experimental science, refined measurements can reveal unexpected behaviour. On the other hand, mathematical extrapolation of existing theories into new theoretical areas, critical reexamination of apparently obvious but untested assumptions, argument by symmetry or analogy, aesthetic judgment, pure accident, and hunch—each of these plays a role (as in all of science). Thus, for example, the quantum hypothesis proposed by the German physicist Max Planck was based on observed departures of the character of blackbody radiation (radiation emitted by a heated body that absorbs all radiant energy incident upon it) from that predicted by classical electromagnetism. The English physicist P.A.M. Dirac predicted the existence of the positron in making a relativistic extension of the quantum theory of the electron. The elusive neutrino, without mass or charge, was hypothesized by the German physicist Wolfgang Pauli as an alternative to abandoning the conservation laws in the beta-decay process. Maxwell conjectured that if changing magnetic fields create electric fields (which was known to be so), then changing electric fields might create magnetic fields, leading him to the electromagnetic theory of light. Albert Einstein's special theory of relativity was based on a critical reexamination of the meaning of simultaneity, while his general theory of relativity rests on the equivalence of inertial and gravitational mass.

Although the tactics may vary from problem to problem, the physicist invariably tries to make un-solved problems more tractable by constructing a series of idealized models, with each successive model being a more realistic representation of the actual physical situation. Thus, in the theory of gases, the molecules are at first imagined to be particles that are as structureless as billiard balls with vanishingly small dimensions. This ideal picture is then improved on step by step.

The correspondence principle, a useful guiding principle for extending theoretical interpretations, was formulated by the Danish physicist Niels Bohr in the context of the quantum theory. It as-serts that when a valid theory is generalized to a broader arena, the new theory's predictions must agree with the old one in the overlapping region in which both are applicable. For example, the more comprehensive theory of physical optics must yield the same result as the more restrictive theory of ray optics whenever wave effects proportional to the wavelength of light are negligible on account of the smallness of that wavelength. Similarly, quantum mechanics must yield the same results as classical mechanics in circumstances when Planck's constant can be considered as neg-ligibly small. Likewise, for speeds small compared to the speed of light (as for baseballs in play), relativistic mechanicsmust coincide with Newtonian classical mechanics.

Some ways in which experimental and theoretical physicists attack their problems are illustrated by the following examples.

The modern experimental study of elementary particles began with the detection of new types of unstable particles produced in the atmosphere by primary radiation, the latter consisting mainly of high-energy protons arriving from space. The new particles were detected in Geiger counters and identified by the tracks they left in instruments called cloud chambers and in photographic plates. After World War II, particle physics, then known as high-energy nuclear physics, became a major field of science. Today's high-energy particle accelerators can be several kilometres in length, cost hundreds (or even thousands) of millions of dollars, and accelerate particles to enormous energies (trillions of electron volts). Experimental teams, such as those that discovered the W+, W−, and Z quanta of the weak force at the European Laboratory for Particle Physics (CERN) in Geneva, which is funded by its 20 European member states, can have 100 or more physicists from many coun-tries, along with a larger number of technical workers serving as support personnel. A variety of visual and electronic techniques are used to interpret and sort the huge amounts of data produced by their efforts, and particle-physics laboratories are major users of the most advanced technology, be it superconductive magnets or supercomputers.

Theoretical physicists use mathematics both as a logical tool for the development of theory and for calculating predictions of the theory to be compared with experiment. Newton, for one, invented integral calculus to solve the following problem, which was essential to his formulation of the law of universal gravitation: Assuming that the attractive force between any pair of point particles is inversely proportional to the square of the distance separating them, how does a spherical distri-bution of particles, such as the Earth, attract another nearby object? Integral calculus, a procedure for summing many small contributions, yields the simple solution that the Earth itself acts as a point particle with all its mass concentrated at the centre. In modern physics, Dirac predicted the existence of the then-unknown positive electron (or positron) by finding an equation for the electron that would combine quantum mechanics and the special theory of relativity.

Physical Quantities and Units

The range of objects and phenomena studied in physics is immense. From the incredibly short lifetime of a nucleus to the age of the Earth, from the tiny sizes of sub-nuclear particles to the vast distance to the edges of the known universe, from the force exerted by a jumping flea to the force between Earth and the Sun, there are enough factors of 10 to challenge the imagination of even the most experienced scientist. Giving numerical values for physical quantities and equations for physical principles allows us to understand nature much more deeply than does qualitative description alone. To comprehend these vast ranges, we must also have accepted units in which to express them. And we shall find that (even in the potentially mundane discussion of meters, kilograms, and seconds) a profound simplicity of nature appears—all physical quantities can be expressed as combinations of only four fundamental physical quantities: length, mass, time, and electric current.

We define a physical quantity either by specifying how it is measured or by stating how it is calculated from other measurements. For example, we define distance and time by specifying methods for measuring them, whereas we define average speed by stating that it is calculated as distance traveled divided by time of travel.

Measurements of physical quantities are expressed in terms of units, which are standardized values. For example, the length of a race, which is a physical quantity, can be expressed in units of meters (for sprinters) or kilometers (for distance runners). Without standardized units, it would be extremely difficult for scientists to express and compare measured values in a meaningful way.

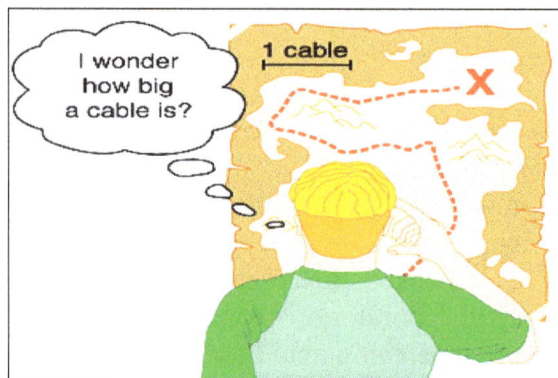

Figure: Distances given in unknown units are maddeningly useless.

There are two major systems of units used in the world: SI units (also known as the metric system) and English units (also known as the customary or imperial system). English units were historically used in nations once ruled by the British Empire and are still widely used in the United States. Virtually every other country in the world now uses SI units as the standard; the metric system is also the standard system agreed upon by scientists and mathematicians. The acronym "SI" is derived from the French Système International.

SI base Quantities and Units

The value of a physical quantity is usually expressed as the product of a *number* and a *unit*. The *unit* represents a specific example or prototype of the quantity concerned, which is used as a point of reference. The *number* represents the ratio of the value of the quantity to the unit. In the case of the *kilogram*, the prototype is a platinum-iridium cylinder held under tightly controlled conditions in a vault at the BIPM, although there are a number of identical copies kept under identical conditions located throughout the world. A quantity of *two kilograms* (2 kg) would have exactly twice the mass of the prototype or one of its copies. There are seven base quantities used in the International System of Units. The seven base quantities and their corresponding units are:

- Length (metre)
- Mass (kilogram)
- Time (second)
- Electric current (ampere)
- Thermodynamic temperature (kelvin)
- Amount of substance (mole)
- Luminous intensity (candela)

These base quantities are assumed to be *independent* of one another. In other words, no base quantity needs to be defined in terms of any other base quantity (or quantities). Note however that although the base quantities themselves are considered to be independent, their respective *base units* are in some cases dependent on one another. The *metre*, for example, is defined as the length of the path travelled by light in a vacuum in a time interval of 1/299 792 458 of a *second*. The table below summarises the base quantities and their units. You may have noticed that an anomaly arises with respect to the *kilogram* (the unit of *mass*). The kilogram is the only SI base unit whose name and symbol include a prefix. You should be aware that multiples and submultiples of this unit are formed by attaching the appropriate prefix name to the unit name *gram*, and the appropriate prefix symbol to the unit symbol *g*. For example, one millionth of a kilogram is one *milligram* (1 mg), and not one *microkilogram* (1 μkg).

SI Base Units				
Qty	**Sym.**	**Unit**	**Unit Sym.**	**Unit Definition**
length	*l*	metre	m	The length of the path travelled by light in 1/299 792 458 of a second
mass	*m*	kilogram	kg	The mass of the International Prototype Kilogram
time	*t*	second	s	The duration of 9 192 631 770 periods of the radiation corresponding to the transition between the two hyperfine levels of the ground state of the caesium 133 atom, at rest, at a temperature of 0 K
electric current	*I*	ampere	A	The constant current which, if maintained in two straight parallel conductors of infinite length, of negligible circular cross-section, and placed one metre apart in a vacuum, would produce between these conductors a force equal to 2 x 10^{-7} newton per metre of length

thermody-namic tempera-ture	T	kelvin	K	The fraction 1/273.16 of the thermodynamic temperature of the triple point of water
amount of substance	n	mole	mol	The amount of substance of a system which contains as many elementary entities as there are atoms in 0.012 kilogram of carbon 12 (elementary entities, which must be specified, may be atoms, molecules, ions, electrons, other particles or specified groups of such particles)
luminous intensity	I_v	candela	cd	The luminous intensity, in a given direction, of a source that emits monochromatic radiation of frequency 540×10^{12} hertz and that has a radiant intensity in that direction of 1/683 watts per steridian

Coherent Derived Units Expressed in Terms of SI base Units

The derived units of quantity identified by the International System of Units are all defined as products of powers of base units. A derived quantity can therefore be expressed in terms of one or more base quantity in the form of an algebraic expression. Derived units that are products of powers of base units that include no numerical factor other than one are said to be coherent derived units. This means that they are derived purely using products or quotients of integer powers of base quantities, and that no numerical factor other than one is involved. The table below gives some examples of coherent derived units.

SI Derived Units expressed in terms of Base Units			
Qty	Sym.	Unit	Unit Sym.
area	A	square metre	m²
volume	V	cubic metre	m³
speed, velocity	v	metre per second	m/s
acceleration	a	metre per second squared	m/s²
wavenumber	σ	reciprocal metre	m⁻¹
density, mass density	ρ	kilogram per cubic metre	kg/m³
surface density	ρ_A	kilogram per square metre	kg/m²
specific volume	v	cubic metre per kilogram	m³/kg
current density	j	ampere per square metre	A/m²
magnetic field strength	H	ampere per metre	A/m
amount concentration, concentration	c	mole per cubic metre	mol/m³
mass concentration	ρ	kilogram per cubic metre	kg/m³
luminance	L_v	candela per square metre	cd/m²
refractive index	n	one	1
relative permeability	μ_r	one	1

Coherent Derived Units with Special Names and Symbols

There are a number of derived SI units that have special names and symbols. Often, the name chosen acknowledges the contribution of a particular scientist. The unit of force (the newton) is named after Sir Isaac Newton, one of the greatest contributors in the field of classical mechanics. The unit of pressure (the pascal) is named after Blaise Pascal for his work in the fields of hydrodynamics and hydrostatics. Each unit named in the table below has its own symbol, but can be defined in terms of other derived units or in terms of the SI base units, as shown in the last two columns.

The units for the plane angle and the solid angle (the radian and steradian respectively) are both derived as the quotient of two identical SI base units. They are thus said to have the unit one (1). They are described as dimensionless units or units of dimension one (the concept of dimension will be described shortly). Note that a temperature difference of one degree Celsius has exactly the same value as a temperature difference of one kelvin. The Celsius temperature scale tends to be used for day-to-day non-scientific purposes such as reporting the weather, or for specifying the temperature at which foodstuffs and medicines should be stored. In this kind of context it is somewhat more meaningful to a member of the public than the Kelvin temperature scale.

SI Derived Units with special names				
Qty	Unit	Unit Sym.	Other Units	Base Units
plane angle	radian	rad	1	m/m
solid angle	steradian	sr	1	m^2/m^2
frequency	hertz	Hz	-	s^{-1}
force	newton	N	-	$m\ kg\ s^{-2}$
pressure, stress	pascal	Pa	N/m^2	$m^{-1}\ kg\ s^{-2}$
energy, work, amount of heat	joule	J	N m	$m^2\ kg\ s^{-2}$
power, radiant flux	watt	W	J/s	$m^2\ kg\ s^{-3}$
electric charge, amount of electricity	coulomb	C	-	s A
electric potential difference, electromotive force	volt	V	W/A	$m^2\ kg\ s^{-3}\ A^{-1}$
capacitance	farad	F	C/V	$m^{-2}\ kg^{-1}\ s^4\ A^2$
electric resistance	ohm	Ω	V/A	$m^2\ kg\ s^{-3}\ A^{-2}$
electric conductance	siemens	S	A/V	$m^{-2}\ kg^{-1}\ s^3\ A^2$
magnetic flux	weber	Wb	V s	$m^2\ kg\ s^{-2}\ A^{-1}$
magnetic flux density	tesla	T	Wb/m^2	$kg\ s^{-2}\ A^{-1}$
inductance	henry	H	Wb/A	$m^2\ kg\ s^{-2}\ A^{-2}$
Celsius temperature	degree Celsius	°C	-	K
luminous flux	lumen	lm	cd sr	cd
illuminance	lux	lx	lm/m^2	$m^{-2}\ cd$
activity referred to a radio nuclide	becquerel	Bq	-	s^{-1}
absorbed dose, specific energy (imparted), kerma	gray	Gy	J/kg	$m^2\ s^{-2}$
dose equivalent, ambient dose equivalent, directional dose equivalent, personal dose equivalent	sievert	Sv	J/kg	$m^2\ s^{-2}$
catalytic activity	katal	kat	-	$s^{-1}\ mol$

Multiples and Submultiples of SI Units

Multiples and submultiples of SI units are signified by attaching the appropriate prefix to the unit symbol. Prefixes are printed as roman (upright) characters prepended to the unit symbol with no intervening space. Most unit multiple prefixes are upper case characters (the exceptions are deca (da), hecto (h) and kilo (k). All unit submultiple prefixes are lower case characters. Prefix names are always printed in lower case characters, except where they appear at the beginning of a sentence, and prefixed units appear as single words (e.g. millimeter, micropascal and so on). All multiples and submultiples are integer powers of ten. Beyond one hundred (or one hundredth) multiples and submultiples are integer powers of one thousand, although they are still expressed as powers of ten. The following table lists the most commonly encountered multiple and submultiple prefixes.

SI Prefixes					
Factor	**Name**	**Symbol**	**Factor**	**Name**	**Symbol**
10^1	deca	da	10^{-1}	deci	d
10^2	hecto	h	10^{-2}	centi	c
10^3	kilo	k	10^{-3}	milli	m
10^6	mega	M	10^{-6}	micro	μ
10^9	giga	G	10^{-9}	nano	n
10^{12}	tera	T	10^{-12}	pico	p
10^{15}	peta	P	10^{-15}	femto	f
10^{18}	exa	E	10^{-18}	atto	a
10^{21}	zetta	Z	10^{-21}	zepto	z
10^{24}	yotta	Y	10^{-24}	yocto	y

Significant Figures

Significant figures are all of the digits that can be known with certainty in a measurement plus an estimated last digit. Significant figures provide a system to keep track of the limits of the original measurement. To record a measurement, you must write down all the digits actually measured, including measurements of zero, and you must not write down any digit not measured. The only real difficulty with this system is that zeros are sometimes used as measured digits, while other times they are used to locate the decimal point.

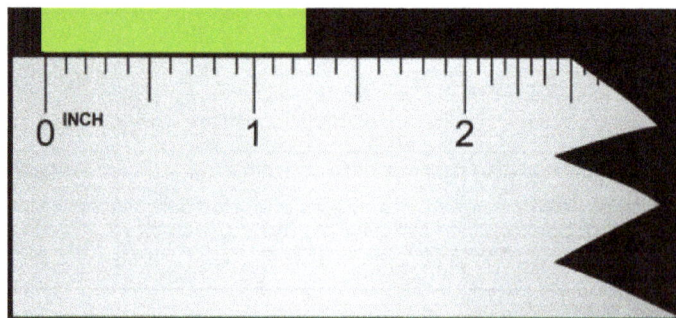

In the figure shown above, the correct measurement is greater than 1.2 inches but less than 1.3 inches. It is proper to estimate one place beyond the calibrations of the measuring instrument.

This ruler is calibrated to 0.1 inches, so we can estimate the hundredths place. This reading should be reported as 1.25 or 1.26 inches.

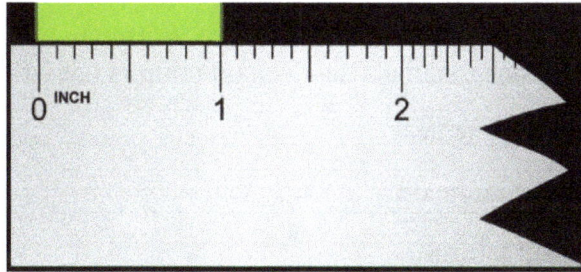

In this second case, it is apparent that the object is, as nearly as we can read, 1 inch. Since we know the tenths place is zero and can estimate the hundredths place to be zero, the measurement should be reported as 1.00 inch. It is vital that you include the zeros in your reported measurement because these are measured places and are significant figures.

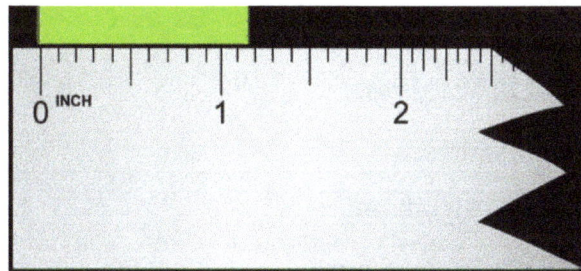

This measurement is read as 1.15 or 1.16 inches.

This measurement is read as 1.50 inches.

In all of these examples, the measurements indicate that the measuring instrument had subdivisions of a tenth of an inch and that the hundredths place is estimated. There is some uncertainty about the last, and only the last, digit.

In our system of writing measurements to show significant figures, we must distinguish between measured zeros and place-holding zeros. Here are the rules for determining the number of significant figures in a measurement.

Rules for Determining the Number of Significant Figures

- All non-zero digits are significant.

- All zeros between non-zero digits are significant.

- All beginning zeros are *not* significant.

- Ending zeros are significant if the decimal point is written in but *not* significant if the decimal point is an understood decimal (the decimal point is not written in).

Examples of the Significant Figure Rules:

- All non-zero digits are significant.

  - 543 has 3 significant figures.

  - 22.437 has 5 significant figures.

  - 1.321754 has 7 significant figures.

- All zeros between non-zero digits are significant.

  - 7,004 has 4 significant figures.

  - 10.3002 has 6 significant figures.

  - 103 has 3 significant figures.

- All beginning zeros are *not* significant.

  - 0.00000075 has 2 significant figures.

  - 0.02 has 1 significant figure.

  - 0.003003 has 4 significant figures.

- Ending zeros are significant if the decimal point is actually written in but *not* significant if the decimal point is an understood decimal.

  - 37.300 has 5 significant figures.

  - 33.00000 has 7 significant figures.

  - 100. has 3 significant figures.

  - 100 has 1 significant figure.

  - 302,000 has 3 significant figures.

  - 1,050 has 3 significant figures.

Addition and Subtraction

The answer to an addition or subtraction operation must not have any digits further to the right than the shortest addend. In other words, the answer should have as many decimal places as the addend with the smallest number of decimal places.

Example:

> 13.3843 cm
> 1.012 cm
> +3.22 cm
> ───────────
> 17.6163 cm = 17.62 cm

The top addend has a 3 in the last column on the right, but neither of the other two addends have a number in that column. In elementary math classes, you were taught that these blank spaces can be filled in with zeros and the answer would be 17.6163 cm. In the sciences, however, these blank spaces are unknown numbers, *not* zeros. Since they are unknown numbers, you cannot substitute any numbers into the blank spaces. As a result, you cannot know the sum of adding (or subtracting) any column of numbers that contain an unknown number. When you add the columns of numbers in the example above, you can only be certain of the sums for the columns with known numbers in each space in the column. In science, the process is to add the numbers in the normal mathematical process and then round off all columns that contain an unknown number (a blank space). Therefore, the correct answer for the example above is 17.62 cm and has only four significant figures.

Multiplication and Division

The answer for a multiplication or division operation must have the same number of significant figures as the factor with the least number of significant figures.

Example

$$(3.556 \text{ cm} \ (2.4 \text{ cm}) = 8.5344 \text{ cm}^2 = 8.5 \text{ cm}^2$$

The factor 3.556 cm has four significant figures, and the factor 2.4 cm has two significant figures. Therefore the answer must have two significant figures. The mathematical answer of 8.5344 cm² must be rounded back to 8.5 cm² in order for the answer to have two significant figures.

Example

$$(20.0 \text{ cm})(5.0000 \text{ cm}) = 100.00000 \text{ cm}^2 = 100. \text{ cm}^2$$

The factor 20.0 cm has three significant figures, and the factor 5.0000 cm has five significant figures. The answer must be rounded to three significant figures. Therefore, the decimal must be written in to show that the two ending zeros are significant. If the decimal is omitted (left as an understood decimal), the two zeros will not be significant and the answer will be wrong.

Example

$$(5.444 \text{ cm})(22 \text{ cm}) = 119.768 \text{ cm}^2 = 120 \text{ cm}^2$$

In this case, the answer must be rounded back to two significant figures. We cannot have a decimal after the zero in 120 cm² because that would indicate the zero is significant, whereas this answer must have exactly two significant figures.

Dimensions and Dimensional Analysis

The dimension of any physical quantity expresses its dependence on the base quantities as a product of symbols (or powers of symbols) representing the base quantities. In table lists the base quantities and the symbols used for their dimension. For example, a measurement of length is said to have dimension L or L^1, a measurement of mass has dimension M or M^1, and a measurement of time has dimension T or T^1. Like units, dimensions obey the rules of algebra. Thus, area is the product of two lengths and so has dimension L^2, or length squared. Similarly, volume is the product of three lengths and has dimension L^3, or length cubed. Speed has dimension length over time, L/T or LT^{-1}. Volumetric mass density has dimension M/L^3 or ML^{-3}, or mass over length cubed. In general, the dimension of any physical quantity can be written as

$$L^a M^b T^c I^d \Theta^e N^f J^g$$

for some powers a, b, c, d, e, f, and g. We can write the dimensions of a length in this form with a = 1 and the remaining six powers all set equal to zero:

$$L^1 = L^1 M^0 T^0 I^0 \Theta^0 N^0 J^0$$

Any quantity with a dimension that can be written so that all seven powers are zero (that is, its dimension is $L^0 M^0 T^0 I^0 \Theta^0 N^0 J^0$) is called dimensionless (or sometimes "of dimension 1," because anything raised to the zero power is one). Physicists often call dimensionless quantities *pure numbers*.

Table: Base Quantities and Their Dimensions	
Base Quantity	Symbol for Dimension
Length	L
Mass	M
Time	T
Current	I
Thermodynamic Temperature	Θ
Amount of Substance	N
Luminous Intensity	J

Physicists often use square brackets around the symbol for a physical quantity to represent the dimensions of that quantity. For example, if r is the radius of a cylinder and h is its height, then we write [r] = L and [h] = L to indicate the dimensions of the radius and height are both those of length, or L. Similarly, if we use the symbol A for the surface area of a cylinder and V for its volume, then [A] = L^2 and [V] = L^3. If we use the symbol m for the mass of the cylinder and $\rho\rho$ for the density of the material from which the cylinder is made, then [m] = M and [$\rho\rho$] = ML^{-3}.

Dimensional Analysis

Dimensional analysis is the use of dimensions and the dimensional formula of physical quantities to find interrelations between them. It is based on the following facts:

Physical Laws

The physical laws are independent of the units in which a quantity is measured. If n1a1 is the measured value of a physical quantity in one system of units and n2a2 is the value in another system of units then, from the above reasoning, these two must be equal.

$$n_1 a_1 = n_2 a_2$$

Principle of Homogeneity

The equations depicting physical situations must have the same dimensions throughout. If two sides of an equation have different dimensions, that equation can't represent any physical situation. This is known as the Principle of Homogeneity. For example, if

$$[M]^a \, [L]^b \, [T]^c = [M]^x \, [L]^y \, [T]^z$$

then from the principle of Homogeneity, we have:

$$a = x;\ b = y;\ c = z$$

Applications of the Dimensional Analaysis

Conversion of Units

The dimensions of a physical quantity are independent of the system of units used to measure the quantity in. Let us suppose that M_1, L_1 and T_1 and M_2, L_2 and T_2 are the fundamental quantities in two different systems of units. We will measure a quantity Q (say) in both these systems of units. Suppose, a, b, c be the dimensions of the quantity respectively.

In the first system of units, $Q = n_1 u_1 = n_1 \, [M_1^a \, L_1^b \, T_1^c]$

In the second system of units, $Q = n_2 u_2 = n_1 \, [M_2^a \, L_2^b \, T_2^c]$

$$n_1 \, [M_1^a \, L_1^b \, T_1^c] = n_2 [M_2^a \, L_2^b \, T_2^c]$$

Substitution of the respective values will give the value of n_1 or n_2

Checking the Consistency of an Equation

All the physical equations must be consistent. The reverse may not be true. For example, the following equations are not consistent because of the dimensions of the L.H.S. ≠ Dimensions of the R.H.S.

$$F = m^2 \times a$$

$$F = m \times a^2$$

Finding relations between physical quantities in a physical phenomenon.

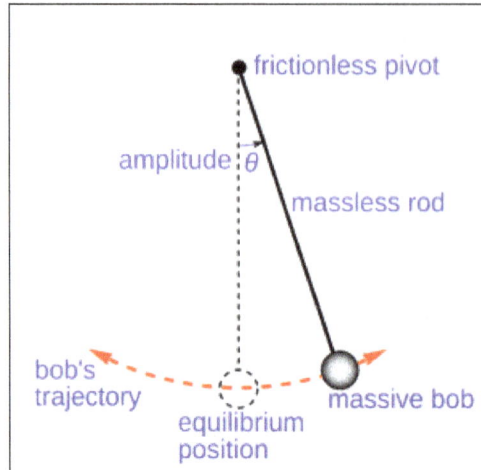

The principle of Homogeneity can be used to derive the relations between various physical quantities in a physical phenomenon.

Solved Example for you

Example: The Time period (T) of a simple pendulum is observed to depend on the following factors:

- Length of the pendulum (L),

- Mass of the Bob (m)

- Acceleration due to gravity (g)

Solution: Let $T \sim L^{\alpha} m^{\beta} g^{\gamma}$ or

$$[T] = [L]^{\alpha} [M]^{\beta} [L]^{\gamma} [T]^{-2\gamma}$$

Solving the above for α, β and γ we have:

$\beta = 0; \alpha + \gamma = 0;$

$-2\gamma = 1$ or $\gamma = -1/2$

Using these in equation $[T] = [L]^{\alpha} [M]^{\beta} [L]^{\gamma} [T]^{-2\gamma}$

we have:

$$T \sim \sqrt{L/g}$$

Example: Convert 1 J to erg.

Solution: Joule is the S.I. unit of work. Let this be the first system if units. Also erg is the unit of work in the cgs system of units. This will be the second system of units. Also n_1 = 1J and we have to find the value of n_2

From equation $n_1 [M_1^a L_1^b T_1^c] = n_2 [M_2^a L_2^b T_2^c]$, we have:

$$[M_1^a L_1^b T_1^c] = n_2 [M_2^a L_2^b T_2^c]$$

The dimensional formula of work is $[M^1 L^2 T^{-2}]$. As a result a = 1, b = 2 and c = -2.

$$[M_1^1 L_1^2 T_1^{-2}] = n_2 [M_2^1 L_2^2 T_2^{-2}]$$

$M_1 = 1 \text{kg}, L_1 = 1 \text{m}, T_1 = 1 \text{s}$

and $M_2 = 1 \text{g}, L_2 = 1 \text{cm}, T_2 = 1 \text{s}$

$n_2 = [M_1^1 L_1^2 T_1^{-2}] / [M_2^1 L_2^2 T_2^{-2}]$

$= \left[1 kg^1 1 g^2 1 s^{-2} \right] / \left[1 g^1 1 cm^2 1 s^{-2} \right]$

$n_2 = 10^7$

Hence, $1 \text{J} = 10^7$ erg

Reference

- Physics-science, science: britannica.com, Retrieved 5 February, 2019

- Physical-quantities-and-units, physics: lumenlearning.com, Retrieved 12 April, 2019

- Physical-quantities-and-si-units, measurement-and-units, physics: technologyuk.net, Retrieved 7 January, 2019

- Dimensional_Analysis, Units_and_Measurement, Waves, Oscillations, Sound, Mechanics: libretexts.org, Retrieved 19 May, 2019

- Dimensional-analysis-applications, units-and-measurement, physics: toppr.com, Retrieved 21 March, 2019

2

Classical Mechanics

Classical mechanics is the study of motion of macroscopic objects such as stars, planets, galaxies and projectiles. Kinematics, dynamics and statics are the three main branches that fall under this domain. The topics elaborated in this chapter will help in gaining a better perspective about classical mechanics.

Classical mechanics is a branch of physics that deals with the motion of bodies based on Isaac Newton's laws of mechanics. Classical mechanics describes the motion of point masses (infinitesimally small objects) and of rigid bodies (large objects that rotate but cannot change shape). While no objects are truly point masses or perfectly rigid, by approximating them as such, classical mechanics accurately describes the motion of objects from molecules to galaxies. Classical mechanics effectively describes systems in which quantum or relativistic effects are negligible.

Kinematics

Kinematics is the study of the motion of points, objects, and groups of objects without considering the causes of its motion.

Kinematics is the branch of classical mechanics that describes the motion of points, objects and systems of groups of objects, without reference to the causes of motion (i.e., forces). The study of kinematics is often referred to as the "geometry of motion."

Objects are in motion all around us. Everything from a tennis match to a space-probe flyby of the planet Neptune involves motion. When you are resting, your heart moves blood through your veins. Even in inanimate objects there is continuous motion in the vibrations of atoms and molecules. Interesting questions about motion can arise: how long will it take for a space probe to travel to Mars? Where will a football land if thrown at a certain angle? An understanding of motion, however, is also key to understanding other concepts in physics. An understanding of acceleration, for example, is crucial to the study of force.

To describe motion, kinematics studies the trajectories of points, lines and other geometric objects, as well as their differential properties (such as velocity and acceleration). Kinematics is used in astrophysics to describe the motion of celestial bodies and systems; and in mechanical engineering, robotics and biomechanics to describe the motion of systems composed of joined parts (such as an engine, a robotic arm, or the skeleton of the human body).

A formal study of physics begins with kinematics. Kinematic analysis is the process of measuring the kinematic quantities used to describe motion. The study of kinematics can be abstracted into purely mathematical expressions, which can be used to calculate various aspects of motion such as velocity, acceleration, displacement, time, and trajectory.

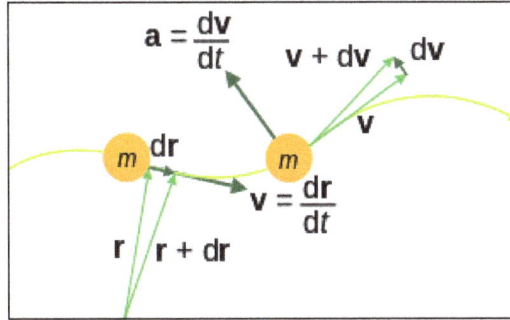

Kinematics of a particle trajectory: Kinematic equations can be used to calculate the trajectory of particles or objects. The physical quantities relevant to the motion of a particle include: mass m, position r, velocity v, acceleration a.

Frame of Reference

Frame of reference – Frame of reference is the set of axes which is used to specify the position of the object in a space. The set of axes is rectangular coordinate system which consist three mutually perpendicular axis X, Y, and Z. The point of intersection of these axes is known as the origin or reference point.

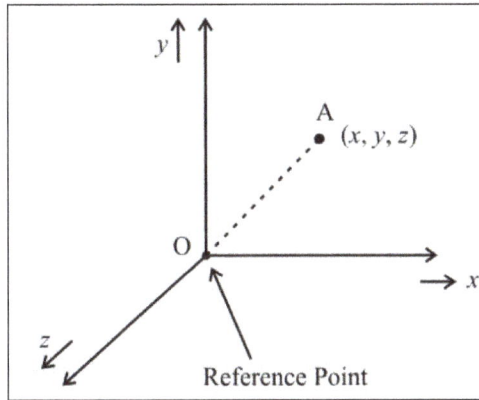

Motion in a Straight Line

Position- Position of any object is essential to describe the motion of the object. The position of object is the set of axes from a reference point.

e.g. In above image the position of point A from the reference point is, $\vec{r} = x\hat{i}\ y\hat{j}\ z\hat{k}$

Motion- An object is said to be in motion if it changes its position with time, with respect to its surroundings. Motion of the object can be represented by the position-time graph. The position-time graph helps to analyze the motion of an object.

Uniform Motion

If an object moving along the straight line covers equal distances in equal interval of time this type of motion is known as uniform motion.

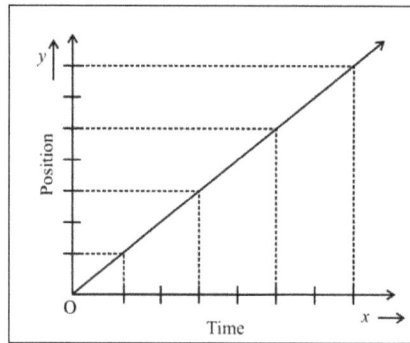

Non-uniform Motion

If an object covers unequal distances in equal interval or equal distance in unequal time interval this type of motion is known as non-uniform motion.

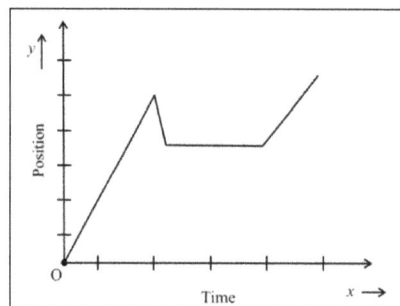

Distance- The length of the actual path between initial and terminal position of a particle in an interval of time is called distance covered by the particle. Distance is also known as the path length.

1. Distance is a scalar quantity.

2. It never reduces with time.

3. Distance of the object can't be negative.

4. SI unit of distance is metre (m).

5. Dimension of the distance is $\left[M^\circ L^1 T^\circ \right]$

Distance time graph- The gradient of distance time graph represents the speed of the object.

Displacement- The difference between the final and initial position is called displacement.

1. Displacement is a vector quantity.

2. Displacement of the object is changes with time.

3. Displacement of the object can be negative, positive or zero.

4. SI unit of displacement is metre (m).

5. Dimension of the distance is $\left[M^\circ L^1 T^\circ \right]$

Displacement time graph- The gradient of displacement time graph represents the velocity.

Speed– Speed of an object is the ratio of distance travelled by the object to the time taken.

$$\text{Speed} = \frac{\text{Distance travelled}}{\text{Time taken}}$$

- Speed is scalar quantity.
- SI unit of speed is m/s.
- Dimension of the speed is $\left[M^{\circ}LT^{-1} \right]$
- Speed of an object can't be negative.

Types of Speed

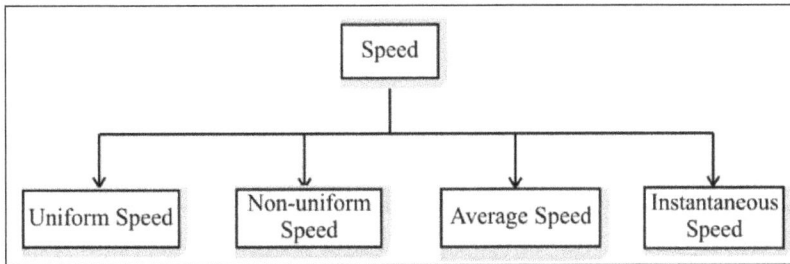

```
                          ┌─────────┐
                          │  Speed  │
                          └─────────┘
                               │
     ┌──────────────┬──────────┴──────────┬──────────────────┐
     ▼              ▼                      ▼                  ▼
┌──────────┐ ┌──────────────┐      ┌──────────────┐ ┌──────────────┐
│ Uniform  │ │ Non-uniform  │      │   Average    │ │Instantaneous │
│  Speed   │ │    Speed     │      │    Speed     │ │    Speed     │
└──────────┘ └──────────────┘      └──────────────┘ └──────────────┘
```

- Uniform speed- An object is said to be moving with a uniform speed, if it covers equal distance in equal intervals of time.

- Non-Uniform speed- An object is said to be non-uniform speed if it covers equal distance in unequal time interval or unequal distance in equal time interval.

- Average speed- The ratio of total path length travelled divided by the total time interval during the motion is known as the average speed of the object.

$$\text{Average speed} = \frac{x_1 + x_2 + x_3 + \ldots}{t_1 + t_2 + t_3 + \ldots} = \frac{\sum\limits_{i=1}^{n} x_i}{\sum\limits_{i=1}^{n} t_i}$$

- Instantaneous speed- The speed of the body at any instant of time or at a position is called instantaneous speed.

$$\text{Instantaneous speed} = \lim_{\Delta t \to 0} \frac{\Delta x}{\Delta t} = \frac{dx}{dt}$$

Velocity- Velocity of an object is the ratio of displacement to the total time taken by object.

$$\text{Velocity} = \frac{\text{Displacement}}{\text{Time}}$$

Velocity is vector quantity.

- SI unit of velocity is m/s.

- Dimension of the velocity is $\left[M^{\circ}LT^{-1} \right]$

- Velocity of an object can be zero, negative, or positive.

Types of Velocity

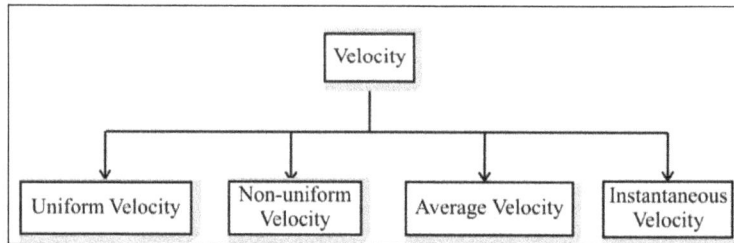

- Uniform velocity- An object is said to be moving with a uniform velocity, if it covers equal distance in equal intervals of time.

- Non-Uniform velocity- An object is said to be non-uniform velocity if it covers equal distance in unequal time interval or unequal distance in equal time interval.

- Average velocity- The ratio of total path length travelled divided by the total time interval during the motion is known as the average velocity of the object.

$$\text{Average velocity} = \frac{x_1 + x_2 + x_3 + ...}{t_1 + t_2 + t_3 + ...} = \frac{\sum_{i=1}^{n} x_i}{\sum_{i=1}^{n} t_i}$$

- Instantaneous velocity- The velocity of the body at any instant of time or at a position is called instantaneous velocity.

$$\text{Instantaneous velocity} = \lim_{\Delta t \to 0} \frac{\Delta x}{\Delta t} = \frac{dx}{dt}$$

Acceleration- The rate of change in velocity of an object is known as the acceleration of the object

$$\text{Acceleration a} = \frac{v_2 - v_1}{t_2 - t_1} = \frac{\Delta v}{\Delta t}$$

- Acceleration is vector quantity.

- SI unit of acceleration is m/s² .

- Dimension of the acceleration is $\left[M^{\circ}LT^{-2} \right]$

- Acceleration of an object can be zero, negative, or positive.

Types of Acceleration

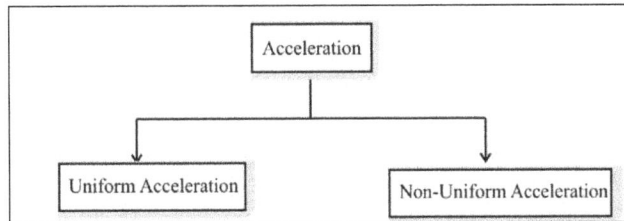

- Uniform Acceleration- A body is said to have uniform acceleration if magnitude and direction of the acceleration both remains constant during motion.

- Non-Uniform Acceleration- A body is said to have non-uniform acceleration if magnitude and direction of the acceleration both change during motion.

Equation of Motion for a Uniformly Accelerated Motion

- $v = u + at$, where v is the final velocity, u is initial velocity, a is the acceleration and t is the time taken during the motion.

- $v^2 = u^2 + 2as$, where v is the final velocity, u is initial velocity, a is the acceleration and s is the distance travelled by object during the motion.

- $s = ut + \frac{1}{2}at^2$, u is initial velocity, a is the acceleration , t is the time taken and s is the distance

- Travelled by object during the motion.

- $S_n = u + \frac{a}{2}(2n-1)$, u is initial velocity, a is the acceleration, s_n is the distance covered by object in nth second.

Equation of Motion for a Free-falling Body Under Gravity

- $v = u + gt$, where v is the final velocity, u is initial velocity, g is the acceleration due to gravity and t is the time taken during the motion.

- $v^2 = u^2 + 2gh$, where v is the final velocity, u is initial velocity, g is the acceleration due to gravity and h is the height covered by object.

- $h = ut + \frac{1}{2}gt^2$, u is initial velocity, g is the acceleration due to gravity, t is the time taken and h is the height covered by object.

- $h_n = u + \frac{g}{2}(2n-1)$, u is initial velocity, g is the acceleration due to gravity, h_n is the height covered by object in nth second.

Relative Velocity

Consider two object X and Y are moving uniformly with velocities v_X and v_Y in one dimension.

Velocity of object Y relative to object X is, $|v_{YX}| = v_Y - v_X$

Velocity of object X relative to object Y is, $|v_{XY}| = v_X - v_Y$

Basic Concept of Vectors

Any physical quantity is classified as vector or scalar.

- Scalar Quantity- Any Physical quantity which can't associated with direction but has magnitude is known as scalar quantity.

- Vector Quantity- Any Physical quantity which has both a direction and a magnitude and obeys triangle law of addition or parallelogram law of addition, is known as vector quantity. A vector quantity is represented in bold or draw an arrow on it.

e.g. A is a vector quantity then A will be represented as $\vec{A}$ or A.

Types of Vector

Unit Vector- Unit vector is a vector that has unit magnitude and points in a particular direction. Unit vector along the x, y, and z axes of a rectangular coordinate system denoted by $\hat{i}, \hat{j},$ and $\hat{k}$ respectively.

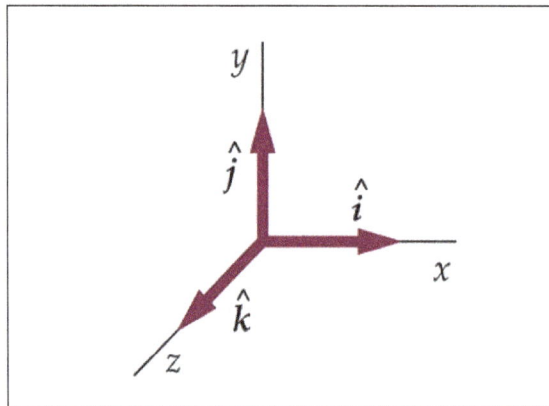

- Zero Vector or null vector- Unit vector is a vector that has zero magnitude. It denoted as $\hat{0}$

- Equal Vector- If two vector A and B have same direction and magnitude then they are equal vector $\vec{A} = \vec{B}$.

- Collinear Vector- Collinear vector are two or more vector which are parallel to the same line irrespective of their magnitude and direction.

Algebra of Vectors

Addition- Let two vectors $\vec{A}$ and $\vec{B}$ to be added. To get the resultant vector the tail of $\vec{}$ coincide with the head of $\vec{A}$. The vector joining the tail of A with the head of B is the vector sum of $\vec{A}$ and $\vec{B}$.

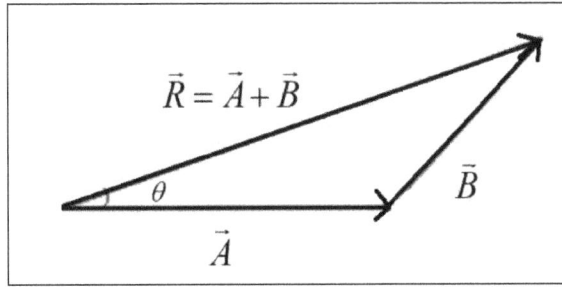

Magnitude of the resultant is, $\mathbf{R} = \sqrt{A^2 + B^2 + 2AB\cos\theta}, \theta$ is the angle between vector $\vec{A}$ and $\vec{B}$.

The vector addition is commutative, $\vec{A} + \vec{B} = \vec{B} + \vec{A}$.

Subtraction- Let two vectors $\vec{A}$ and $\vec{B}$ to be subtracted. Let θ is the angle between vector $\vec{A}$ and $\vec{B}$.

To subtract $\vec{B}$ from $\vec{A}$, invert the direction of $\vec{B}$ and add to $\vec{A}$.

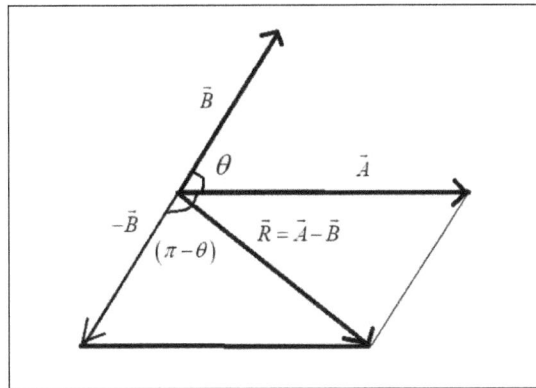

Magnitude of the resultant is,

$$\mathbf{R} = \sqrt{A^2 + B^2 + 2AB\cos(\pi - \theta)}$$
$$\mathbf{R} = \sqrt{A^2 + B^2 - 2AB\sin(\theta)}$$

If we multiply a vector $\vec{A}$ with a positive number X, it gives a vector whose magnitude is changed by the factor X but the direction is the same as that of $\vec{A}$.

$$|X\mathbf{A}| = X|\mathbf{A}|, \text{ if } X > 0$$

Resolution of Vectors

If the vector is not in the X-Y plane, it may have non-zero projections along X, and Y axes and we can resolve it into parts

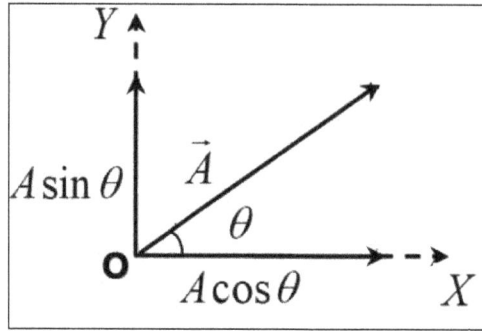

$$\vec{A} = A_x \ \cos\theta + A_y \ \sin\theta$$

Magnitude of vector is $A = \sqrt{A_x^2 + A_y^2}$

Angle between the vector is, $\tan\theta = \dfrac{A_x}{A_a}$

Similarly, we can resolve a vector into three components along X, Y, and Z.

$$\vec{A} = A\cos\alpha + A\cos\beta + A\cos\gamma$$

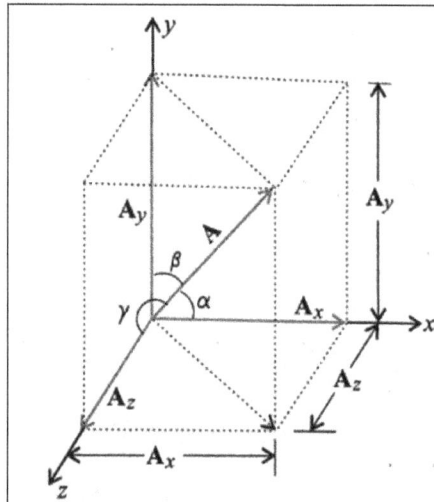

Magnitude of vector is, $A = \sqrt{A_x^2 + A_y^2 + A_z^2}$

Scalar and Vector Product of Vector

The multiplication of vector is two type one is scalar and other is vector.

- Scalar Product

The scalar product or dot product of two vectors $\vec{A}$ and $\vec{B}$ is not a vector, but a scalar quantity.

Let the vector A is, $\vec{A} = A_x \ \hat{i} + A_y \ \hat{j} + A_z \ \hat{k}$ and Vector B is, $\vec{B} = B_x \ \hat{i} + B_y \ \hat{j} + B_z \ \hat{k}$

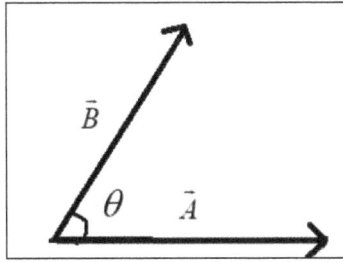

The scalar or dot product of the vector is

$$\vec{A}.\vec{B} = \left(A_x B_x + A_y B_y + A_z B_z \right) \text{ or } \left| \vec{A}.\vec{B} \right| = AB\cos\theta \text{ ,where } \theta \text{ is the angle between vector } \vec{A} \text{ and } \vec{B}$$

The vector product or cross product of two vectors $\vec{A}$ and $\vec{B}$ is not a vector, but a vector quantity.

Let the vector A is, $\vec{A} = A_x \hat{i} + A_y \hat{j} + A_z \hat{k}$ and Vector B is, $\vec{B} = B_x \hat{i} + B_y \hat{j} + B_z \hat{k}$

Then the vector product is, $\left| \vec{A} \times \vec{B} \right| = AB\sin\theta\,\hat{n}, \theta$ is the angle between vector $\vec{A}$ and $\vec{B}$, $\hat{n}$ at right angles to both $\vec{A}$ and $\vec{B}$.

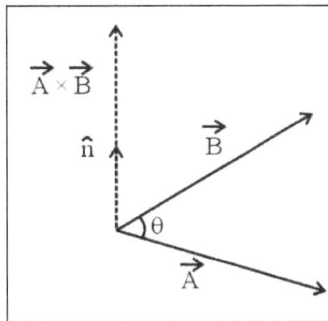

The direction of resultant vector $\vec{A} \times \vec{B}$ is perpendicular to both $\vec{A}$ and $\vec{B}$.

Motion in a Plane

Motion in a Plane with Constant Acceleration

If an object is moving in a two-dimensional plane, then we can treat two separate simultaneous onedimensional motion with constant acceleration along two perpendicular direction.

Consider an object is moving in a two-dimensional plane with velocity $\vec{v}$ and acceleration $\vec{a}$.

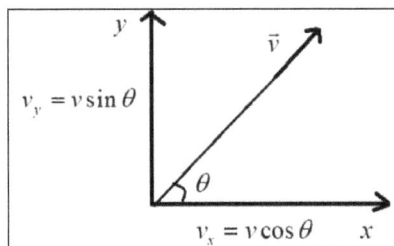

Velocity of the object, $\vec{v} = v_x \hat{i} + v_y \hat{j}$

Acceleration of the object, $\vec{a} = a_x \hat{i} + a_y \hat{j}$

x - axis Motion	y - axis Motion
$v_x = u_x + at$	$v_y = u_y + at$
$s_x = u_x t + \dfrac{1}{2} at^2$	$s_y = u_y t + \dfrac{1}{2} at^2$
$v_x^2 = u_x^2 + 2as_x$	$v_y^2 = u_y^2 + 2as_y$

Relative Velocity in Two Dimensions

Suppose that two objects P and Q are moving uniformly with velocities v_P and v_Q in two-dimensional (x-y) plane. Their velocity $v_P = v_{Px} \hat{i} + v_{Py} \hat{j}$ and $v_Q = v_{Qx} \hat{i} + v_{Qy} \hat{j}$.

Velocity of object Q relative to object P is,

$$\left| v_{QP} \right| = v_Q - v_P$$

$$\left| v_{QP} \right| = \left(v_{Qx} \hat{i} + v_{Qy} \hat{j} \right) - \left(v_{Px} \hat{i} + v_{Py} \hat{j} \right)$$

$$\left| v_{QP} \right| = \left(v_{Qx} - v_{Px} \right) \hat{i} + \left(v_{Qy} - v_{Py} \right) \hat{j}$$

Velocity of object P relative to object Q is,

$$\left| v_{PQ} \right| = v_P - v_Q$$

$$\left| v_{PQ} \right| = \left(v_{Px} \hat{i} + v_{Py} \hat{j} \right) - \left(v_{Qx} \hat{i} + v_{Qy} \hat{j} \right)$$

$$\left| v_{PQ} \right| = \left(v_{Px} - v_{Qx} \right) \hat{i} + \left(v_{Py} - v_{Qy} \right) \hat{j}$$

Projectile Motion

Projectile motion is a motion in which object is moved in a parabolic path. The motion of the object is the result of two separate components of motions. One component is along a horizontal direction without any acceleration and the other along the vertical direction with constant acceleration due to the force of gravity.

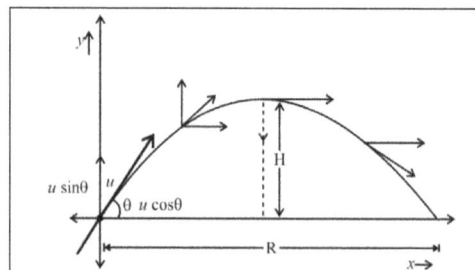

Horizontal Velocity of the particle is, $u_x = u \cos \theta$

Vertical Velocity of the particle is, $u_y = u \cos \theta$

Equation of trajectory is, $y = x \tan \theta - \dfrac{1}{2} g \dfrac{x^2}{u^2} \sec^2 \theta$

Range of the particle is, $R = \dfrac{u^2 \sin 2\theta}{g}$

Time of flight is, $T = \dfrac{2u \sin \theta}{g}$

Height of the projectile is, $H = \dfrac{u^2 \sin^2 \theta}{2g}$

Projectile Motion on an Inclined Plane

Let us assume that a particle is projected from an incline plane which is incline at an angle φ to the horizon. Particle is moving with a velocity u at angle of elevation θ .

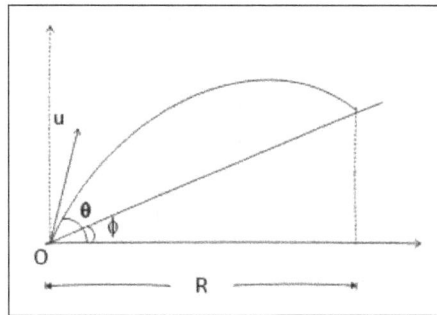

Range of the projectile is, $R = \dfrac{u^2}{g \cos^2 \phi} \left[\sin(2\theta - \phi) - \sin \phi \right]$

Time of flight is, $T = \dfrac{2u \sin(\theta - \phi)}{g \cos \phi}$

Dynamics

Dynamics is the study of the motion of objects (i.e. kinematics) and the forces responsible for that motion. It is a branch of classical mechanics, involving primarily Newton's laws of motion. As a field of study it is very important for analyzing systems consisting of single bodies or multiple bodies interacting with each other.

A dynamics analysis is what allows one to predict the motion of an object or objects, under the influence of different forces, such as gravity or a spring. It can be used to predict the motion of planets in the solar system or the time it takes for a car to brake to a full stop.

Force

A force is a push or pull upon an object resulting from the object's interaction with another object. Whenever there is an interaction between two objects, there is a force upon each of the objects. When the interaction ceases, the two objects no longer experience the force. Forces only exist as a result of an interaction.

Contact versus Action-at-a-Distance Forces

For simplicity sake, all forces (interactions) between objects can be placed into two broad categories:

- Contact forces, and

- Forces resulting from action-at-a-distance

Contact forces are those types of forces that result when the two interacting objects are perceived to be physically contacting each other. Examples of contact forces include frictional forces, tensional forces, normal forces, air resistance forces, and applied forces.

Action-at-a-distance forces are those types of forces that result even when the two interacting objects are not in physical contact with each other, yet are able to exert a push or pull despite their physical separation. Examples of action-at-a-distance forces include gravitational forces. For example, the sun and planets exert a gravitational pull on each other despite their large spatial separation. Even when your feet leave the earth and you are no longer in physical contact with the earth, there is a gravitational pull between you and the Earth. Electric forces are action-at-a-distance forces. For example, the protons in the nucleus of an atom and the electrons outside the nucleus experience an electrical pull towards each other despite their small spatial separation. And magnetic forces are action-at-a-distance forces. For example, two magnets can exert a magnetic pull on each other even when separated by a distance of a few centimeters.

Examples of contact and action-at-distance forces are listed in the table below.

Contact Forces	Action-at-a-Distance Forces
Frictional Force	Gravitational Force
Tension Force	Electrical Force
Normal Force	Magnetic Force
Air Resistance Force	
Applied Force	
Spring Force	

The Newton

Force is a quantity that is measured using the standard metric unit known as the Newton. A Newton is abbreviated by an "N." To say "10.0 N" means 10.0 Newton of force. One Newton is the amount of force required to give a 1-kg mass an acceleration of 1 m/s/s. Thus, the following unit equivalency can be stated:

$$1 \text{ Newton} = 1 \text{ kg} \cdot \text{m}/\text{s}^2$$

Force as a Vector Quantity

A force is a vector quantity. A vector quantity is a quantity that has both magnitude and direction. To fully describe the force acting upon an object, you must describe both the magnitude (size or numerical value) and the direction. Thus, 10 Newton is not a full description of the force acting upon an object. In contrast, 10 Newton, downward is a complete description of the force acting upon an object; both the magnitude (10 Newton) and the direction (downward) are given.

Because a force is a vector that has a direction, it is common to represent forces using diagrams in which a force is represented by an arrow. The size of the arrow is reflective of the magnitude of the force and the direction of the arrow reveals the direction that the force is acting. (Such diagrams are known as free-body diagrams). Furthermore, because forces are vectors, the effect of an individual force upon an object is often canceled by the effect of another force. For example, the effect of a 20-Newton upward force acting upon a book is *canceled* by the effect of a 20-Newton downward force acting upon the book. In such instances, it is said that the two individual forces *balance each other*; there would be no unbalanced force acting upon the book.

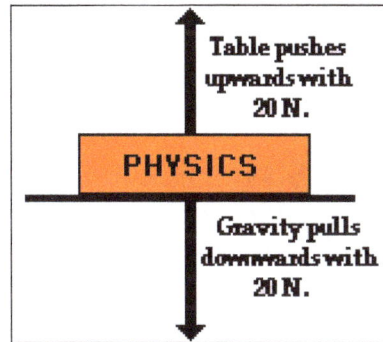

Other situations could be imagined in which two of the individual vector forces cancel each other ("balance"), yet a third individual force exists that is not balanced by another force. For example, imagine a book sliding across the rough surface of a table from left to right. The downward force of gravity and the upward force of the table supporting the book act in opposite directions and thus balance each other. However, the force of friction acts leftwards, and there is no rightward force to balance it. In this case, an unbalanced force acts upon the book to change its state of motion.

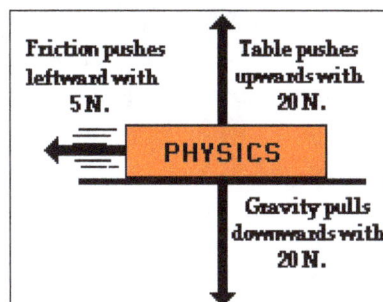

Newton's Laws of Motion

Newton's Laws of Motion help us to understand how objects behave when they are standing still; when they are moving, and when forces act upon them. There are three laws of motion.

Newton's First Law of Motion

Newton's First Law of Motion states that an object in motion tends to stay in motion unless an external force acts upon it. Similarly, if the object is at rest, it will remain at rest unless an unbalanced force acts upon it. Newton's First Law of Motion is also known as the Law of Inertia.

Basically, what Newton's First Law is saying is that objects behave predictably. If a ball is sitting on your table, it isn't going to start rolling or fall off the table unless a force acts upon it to cause it to do so. Moving objects don't change their direction unless a force causes them to move from their path.

As you know, if you slide a block across a table, it eventually stops rather than continuing on forever. This is because the frictional force opposes the continued movement. If you threw a ball out in space, there is much less resistance, so the ball would continue onward for a much greater distance.

Newton's Second Law of Motion

Newton's Second Law of Motion states that when a force acts on an object, it will cause the object to accelerate. The larger the mass of the object, the greater the force will need to be to cause it to accelerate. This Law may be written as force = mass x acceleration or:

$$F = m * a$$

Another way to state the Second Law is to say it takes more force to move a heavy object than it does to move a light object. Simple, right? The law also explains deceleration or slowing down. You can think of deceleration as acceleration with a negative sign on it. For example, a ball rolling down a hill moves faster or accelerates as gravity acts on it in the same direction as the motion (acceleration is positive). If a ball is rolled up a hill, the force of gravity acts on it in the opposite direction of the motion (acceleration is negative or the ball decelerates).

Derivation of Newton's Second Law of Motion

According to Newton's second law: $\vec{F} = d\vec{P} / dt$,

where, $\vec{P}$ = momentum and $\vec{P} = m\vec{v}$

If the time interval for the applied force is increased, then the value of the applied force will decrease. In cricket players use this while catching the ball. They pull their hands back so that time of contact with ball increases and they would experience less jerk due to the motion of the ball.

From Newton's second law of motion,

$$\vec{F} \propto d\vec{P} / dt$$

$$\vec{F} = k \times d\vec{P} / dt = km\vec{a}$$

For simplicity, the constant of proportionality (k) is chosen to be 1, therefore

$$\vec{F} = m\vec{a}$$

Newton's Third Law of Motion

Newton's Third Law of Motion states that for every action, there is an equal and opposite reaction.

What this means is that pushing on an object causes that object to push back against you, the exact same amount, but in the opposite direction. For example, when you are standing on the ground, you are pushing down on the Earth with the same magnitude of force that it is pushing back up at you.

Applications of Laws of Motion

Newton's First Law of Motion

A car travelling on a highway at a fixed speed tends to maintain uniformity in its motion and everything else inside the car. When a force from outside is applied to the car in motion, like a sudden change in direction, the car will respond to this sudden change on its own, although the passengers in the car or the objects inside it are still responding to inertia, wherein their motion will still be in a straight line. When in fact the direction has already changed causing the passengers or the objects to be thrown off. This event is explained by the first law of motion.

Newton's Second Law of Motion

The application of the second law of motion can be seen in determining the amount of force needed to make an object move or to make it stop. For example, stopping a moving ball or pushing a ball.

Newton's Third Law of Motion

The application third law of motion can be seen via an illustration wherein let's say a glass is on a table, even though the glass is at rest it is actually exerting a force on the table and the table, on the other hand, is exerting equal opposite force thus making the glass stay.

Circular Motion

When an object moves in a circle at a constant speed its velocity (which is a vector) is constantly changing. Its velocity is changing not because the magnitude of the velocity is changing but because its direction is. This constantly changing velocity means that the object is accelerating (centripetal acceleration). For this acceleration to happen there must be a resultant force, this force is called the centripetal force.

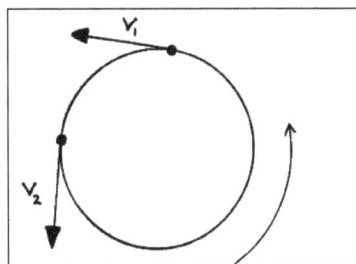

The angular speed (ω) of an object is the angle (θ) it moves through measured in radians (rad) divided by the time (t) taken to move through that angle. This means that the unit for angular speed is the radian per second (rad s^{-1}).

$$\omega = \frac{\theta}{t}$$

v is the linear velocity measured in metres per second (ms^{-1}).

r is the radius of the circle in metres (m).

f is the frequency of the rotation in hertz (Hz).

$$\omega = \frac{v}{r} = 2\pi f$$

Centripetal Acceleration

Centripetal acceleration (a) is measure in metres per second per second (ms^{-2}). It is always directed towards the center of the circle.

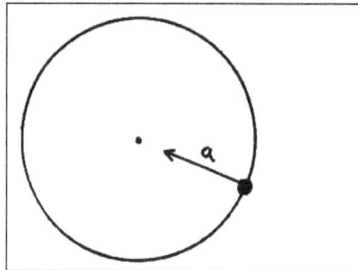

$$a = \frac{v^2}{r} = \omega^2 r$$

Centripetal Force

When an object moves in a circle the centripetal force (F) always acts towards the centre of the circle. The centripetal force, measured in newtons (N) can be different forces in different settings it can be gravity, friction, tension, lift, electrostatic attraction etc.

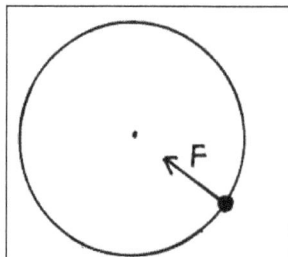

$$F = \frac{mv^2}{r} = m\omega^2 r$$

Rotational Motion

We see rotational motion in almost everything around us. Every machine, celestial bodies, most of the fun games in amusement parks and if you are a FIFA fan, and when you watch the David Beckham's familiar shot, the ball is actually executing rotational motion.

Objects turn about an axis. All the particles and the mass centre do not undergo identical motions. All the particles of the body undergo identical motion. By definition, it becomes essential for us to explore how the different particles of a rigid body move when the body rotates.

Rotational Kinematics

In rotational kinematics, we will investigate the relation between kinematical parameters of rotation. We shall now revisit angular equivalents of the linear quantities: position, displacement, velocity and acceleration which we have already dealt in a circular motion.

Linear Kinematic Parameters	Angular Kinematic Parameters
Position s	Angular position θ
Displacement $\Delta s = s_1 - s_2$	Angular displacement $\Delta \theta = \theta_1 - \theta_2$
Average velocity $v_{avg} = \dfrac{\Delta s}{\Delta t}$	Average angular velocity $\omega_{avg} = \dfrac{\Delta \theta}{\Delta t}$
Instantaneous velocity $v_{ins} \lim_{\Delta t \to 0} \dfrac{\Delta s}{\Delta t} = \dfrac{ds}{dt}$	Instantaneous angular velocity $\omega_{ins} \lim_{\Delta t \to 0} \dfrac{\Delta \theta}{\Delta t} = \dfrac{d\theta}{dt}$
Average acceleration $a_{avg} = \dfrac{\Delta v}{\Delta t}$	Average angular acceleration $\alpha_{avg} = \dfrac{\Delta s}{\Delta t}$
Instantaneous acceleration $a_{ins} \lim_{\Delta t \to 0} \dfrac{\Delta v}{\Delta t} = \dfrac{dv}{dt}$	Instantaneous angular acceleration $\alpha_{ins} \lim_{\Delta t \to 0} \dfrac{\Delta \omega}{\Delta t} = \dfrac{d\omega}{dt}$

A case of constant angular acceleration is of great importance and a parallel set of equations holds for this case just as in constant linear acceleration.

Linear Equations of Motion	Angular Equations of Motion
$v = v_0 + at$	$\omega = \omega_0 + \alpha t$
$x - x_0 = v_0 + \dfrac{1}{2} at^2$	$\theta - \theta_0 = \omega_0 + \dfrac{1}{2} \alpha t^2$
$v^2 = v_0^2 + 2a(x - x_0)$	$\omega^2 = \omega_0^2 + 2\alpha(\theta - \theta_0)$

Axis of Rotation

A rigid body of an arbitrary shape in rotation about a fixed axis (axis that does not move) called axis of rotation or rotation axis is shown in the figure.

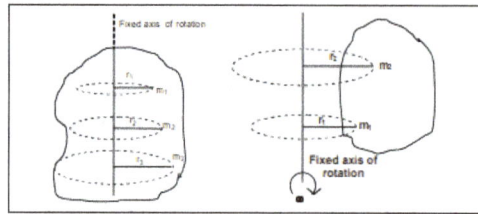

Types of Motion involving Rotation

1. Rotation about a fixed axis (Pure rotation)

2. Rotation about an axis of rotation (Combined translational and rotational motion)

3. Rotation about an axis in the rotation

Rotation about a Fixed Axis

Rotation of a ceiling fan, opening and closing of the door, rotation of our planet, rotation of hour and minute hands in analogue clocks are few examples of this type.

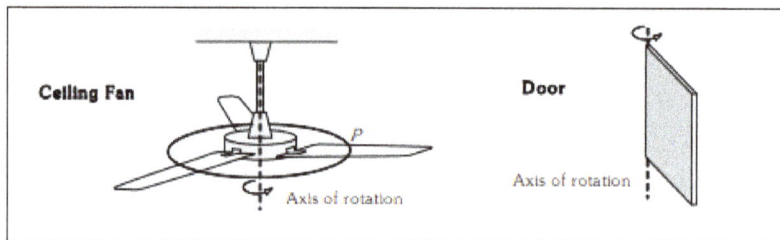

Rotation about an Axis of Rotation

Rolling is an example of this category. Arguably, the most important application of rotational physics is in the rolling of wheels and wheels like objects as our world now is filled with automobiles and other rolling vehicles.

Rolling Motion of a body is a combination of both translational and rotational motion of a round shaped body placed on a surface. When a body is set in rolling motion, every particle of body has two velocities – one due to its rotational motion and the other due to its translational motion (of the centre of mass), and the resulting effect is the vector sum of both velocities at all particles.

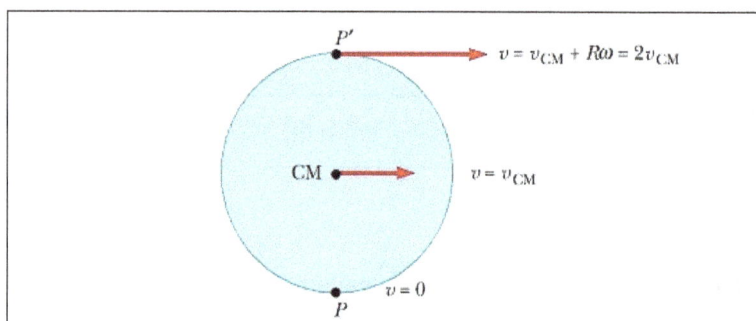

Kinetic Energy of Rotation

The rapidly rotating blades of a table saw machine and the blades of a fan certainly have kinetic energy due tothe rotation. If we apply the familiar equation to the saw machine as a whole, it would give us kinetic energy of its centre of mass only, which is zero.

The right approach:

We shall treat the saw machine or any rotating rigid body as a collection of particles with different speeds. We shall sum up all the kinetic energies of the particles to find the rotational kinetic energy of the whole body.

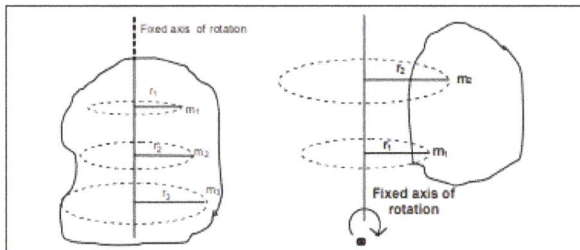

If m_1, m_2, ...m_n are the masses of the constituent particles moving on circular paths with radii r_1, r_2 ... r_n with velocities v_1, v_2, ...v_n, then the kinetic energy of the body is given by

$$KE = \sum_{1}^{n} \frac{1}{2} m_i v_i^2 = \sum_{i}^{n} \frac{1}{2} m_i r_i^2 \omega^2 = \frac{1}{2} \omega^2 \sum_{i}^{n} m_i r_i^2 = \frac{1}{2} I \omega^2$$

The terms $\sum_{i}^{n} m_i r_i^2$ is called rotational inertia or moment of inertia of the system of particles. For a continuous distribution of mass $I = \int_{body} r^2 dm$ where dm is the mass of a particle at a distance r from the axis of rotation.

Torque

Torque is a rotational analogue of force and expresses the tendency of a force applied to an object to cause the object to rotate about a given point.

If you want to open a door, you will apply a force on the doorknob which is located as far as possible from the hinges of the door. If you try to apply the force nearer to the hinge line than the knob, or at any other angle other than 90° to the plane of the door, you must apply greater force than the former to rotate the door.

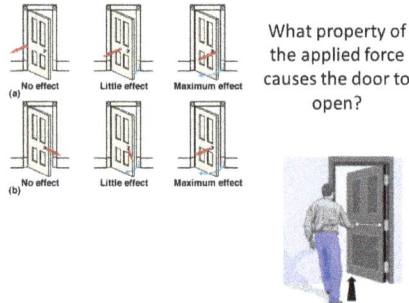

What property of the applied force causes the door to open?

To determine how the applied force results in a rotation of the body about an axis, we resolve the Force (F) into two components. The tangential component (Fsinθ) is perpendicular to r and it does cause rotation whereas the radial component (Fcosθ) does not cause rotation because it acts along the line that intersects with the axis or pivot point.

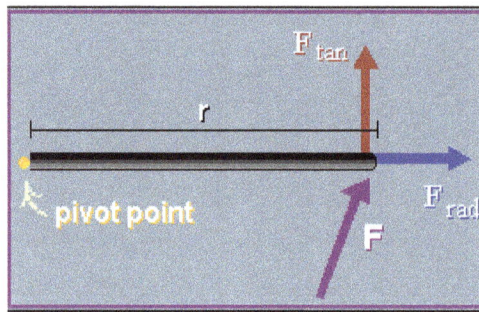

The ability to rotate the body depends on the magnitude of the tangential component and also on how far from axis the force (r – moment of an arm) is applied. Therefore, mathematically it can be represented as $\vec{\tau} = \vec{r} \times \vec{F}$.

SI unit of torque is Nm.

To find the direction of $\vec{\tau}$, we use right hand thumb rule sweeping the fingers from $\vec{r}$ (the first vector in the product) into $\vec{F}$ (the second vector in the product), the outstretched thumb will give the direction of $\vec{\tau}$.

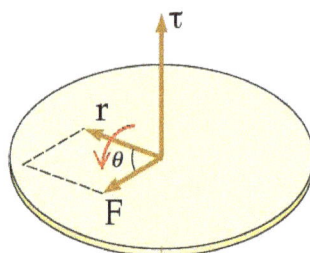

Right-hand rule

Newton's Second Law of Rotation

If the net torque acting on a body about any inertial axis is $\vec{\tau}$ and the moment of inertia about that axis is I, then the angular acceleration of the body is given by the relation:

$$\vec{\tau} = I\vec{\alpha}$$

Rotational Equilibrium

The centre of mass of a body remains in equilibrium if the total external force acting on the body is zero. This follows from the equation F = Ma.

Similarly, a body remains in rotational equilibrium if the total external torque acting on the body is zero. This follows from the equation $\tau = I\alpha$. Therefore a body in rotational equilibrium must either be in rest or rotation with constant angular velocity.

Thus, if a body remains at rest in an inertial frame, the total external force acting on the body should be zero in any direction and the total external torque should be zero about any line.

Under the action of several coplanar forces, the net torque is zero for rotational equilibrium.

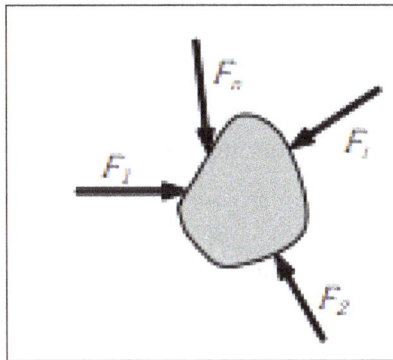

If the net force on the body is zero, then the net torque may or may not be zero.

Ex. Determine the point of application of third force for which body is in equilibrium when forces of 20 N & 30 N are acting on the rod as shown figure.

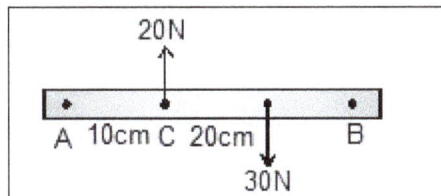

Sol. Let the magnitude of third force is F, is applied in upward direction then the body is in the equilibrium when,

(i) $\vec{F}_{net} = 0$ (Translational Equilibrium)

$$\Rightarrow 20 + F = 30 \Rightarrow F = 10N$$

So the body is in translational equilibrium when 10 N force act on it in upward direction

(ii) Let us assume that this 10N force act.

Then keep the body in rotational equilibrium

So Torque about C = 0

$i.e. = t_c = 0$

$\Rightarrow 30 \times 20 = 10x$

$x = 60\,cm$

So 10 N force is applied at 70 cm from point A to keep the body in equilibrium

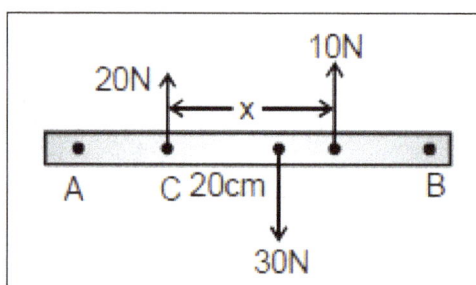

Ex. A stationary uniform rod of mass 'm', length 'l' leans against a smooth vertical wall making an angle q with rough horizontal floor. Find the normal force & frictional force that is exerted by the floor on the rod?

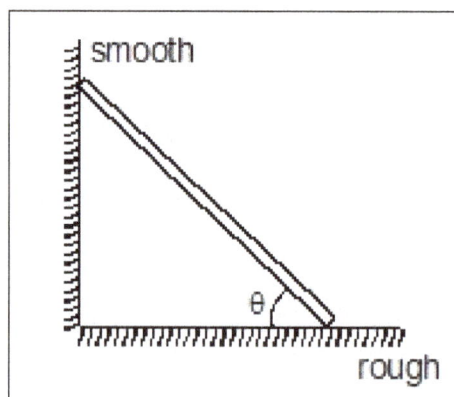

Sol. As the rod is stationary so the linear acceleration and angular acceleration of rod is zero,

$$i.e., a_{cm} = 0 \; ; \alpha = 0.$$

$$\left. \begin{array}{l} N_2 = f \\ N_1 = mg \end{array} \right\} \because a_{cm} = 0$$

Torque about any point of the rod should also be zero,

$$\therefore \alpha = 0$$

$$\tau_A = 0 \Rightarrow mg\ \cos\theta\frac{\ell}{2} + f\ \ell\sin\theta = N_1\cos\theta.\ell$$

$$N_1\cos\theta = \sin\theta f + \frac{mg\cos\theta}{2}$$

$$f = \frac{mg\cos\theta}{2\sin\theta} = \frac{mg\cot\theta}{2}$$

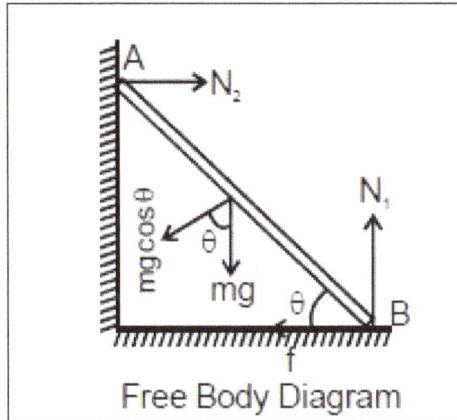

Free Body Diagram

Ex. The ladder shown in figure has negligible mass and rests on frictionless floor. The crossbar connects the two legs of the ladder at the middle. The angle between the two legs is $60°$. The fat person sitting of the ladder has a mass of 80 kg. Find the contanct force exerted by the floor on each leg and the tension in the crossbar.

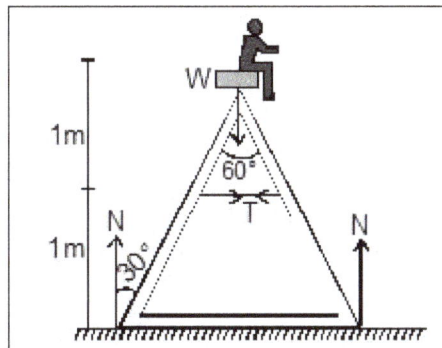

Sol. The forces acting on different parts are shown in figure. Consider the vertical equilibrium of "the ladder plus the person" system. The forces acting on this system are its weight (80 kg)g and the contact force N + N = 2 N due to the floor. Thus

$$2N = (80kg)g \ or \ N = (40kg)(9.8m/s^2) = 392\ N$$

Next consider the equilibrium of the left leg of the ladder. Taking torques of the forces acting on it about the upper end,

$$N(2m)\tan 30° \ T = (1m) \quad or \ T = N\frac{2}{\sqrt{3}} = (392\,N)\times\frac{2}{\sqrt{3}} = 450\,N$$

Ex. The pulley shown in figure has moments of inertia I about its axis and radius R. Find the acceleration of the two blocks. Assume that the string is light and does not slip on the Pulley.

Sol. Suppose the tension in the left string is T_1 and that in the right string is T_2. Suppose the block of mass M goes down with an acceleration a and the other block moves up with the same acceleration. This is also the tangential acceleration of the rim of the wheel as the string does not slip over the rim.

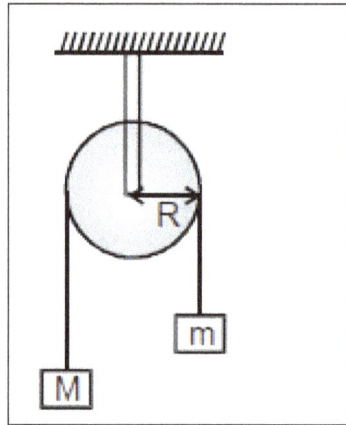

The angular acceleration of the wheel $\alpha = \dfrac{a}{R}$.

The equations of motion for the mass M,

The mass m and the pulley are as follows;

$$Mg - T_1 = Ma \qquad ..(i)$$

$$T_2 - mg = ma \qquad ...(ii)$$

$$T_1 R - T_2 R = I\alpha = \frac{Ia}{R} \quad ...(iii)$$

Substituting for T_1 and T_2 from equations (i) and (ii) in equation (iii)

$$\left[M(g-a) - m(g+a)\right]R = \frac{Ia}{R}$$

Solving, we get

$$a = \frac{(M-m)gR^2}{I + (M+m)R^2}$$

Angular Momentum

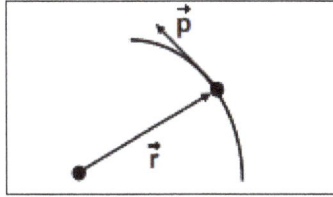

The concept of linear momentum and conservation of linear momentum are extremely powerful tools to predict the collision of two objects without any other details of collision. Thus, the angular counterpart, angular momentum plays a crucial role in orbital mechanics.

Angular momentum of a particle about a given point is given by,

$$\vec{l} = \vec{r} \times \vec{p} = m\left(\vec{r} \times \vec{v}\right)$$

The direction of angular momentum is also given by right hand rule. (refer torque)

Newton's Law in Angular form

The vector sum of all the torques acting on a particle is equal to the time rate of change of the,

$$\vec{\tau}_{net} = \frac{d\vec{t}}{dt}.$$

Angular momentum of a particle describing circular motion:

$\vec{L} = \vec{r} \times \vec{p}$; as linear momentum $\left(\vec{p}\right)$ is along the tangent, hence

$\vec{r} \times \vec{p} = rp\hat{n}$, where $\hat{n}$ is the unit vector perpendicular to the plane of the circle,

$$\Rightarrow \left|\vec{L}\right| = mvr = m\omega r^2, \text{ where } \left|\vec{p}\right| = mv = m\omega r$$

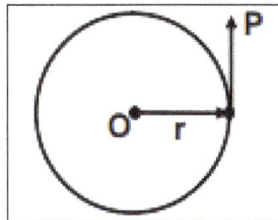

Angular Momentum of a Rigid Body in a Fixed Axis Rotation

In a fixed axis rotation, all the constituent particles describes circular motion.

Hence, angular momentum of the particles about corresponding centres are,

$$L_1 = m_1 \omega r_1^2, \ L_2 = m_2 \omega r_2^2$$

And similarly $L_1 = m_n \omega r_n^2$ where ω is the angular speed of the body.

Since angular momentum of all the particles have same direction therefore angular momentum of the whole body is given by,

$$L = L_1 + L_2 + \ldots\ldots\ldots\ldots L_n$$
$$= \left(m_1 r_1^2 + m_2 r_2^2 + \ldots\ldots + m_n r_n^2 \right) \omega$$
$$\Rightarrow L = I\omega$$

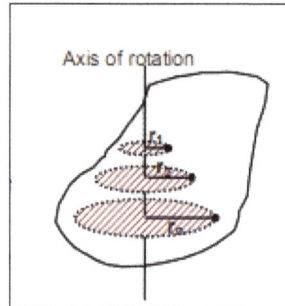

Axis of rotation

Conservation of Angular Momentum

By the definition of torque,

$$\vec{\tau}_{net} = \frac{d\vec{l}}{dt}, if\ \vec{\tau}_{net} = 0,\ then\ \frac{d\vec{l}}{dt} = 0, \vec{l} = \text{constant}$$

When the resultant torque acting on a system is zero, then the total vector angular momentum of the system remains constant. This is called as principle of conservation of angular momentum.

Energy

In Physics, energy is the capacity for doing work. It may exist in potential, kinetic, thermal, electrical, chemical, nuclear, or other various forms. There are, moreover, heat and work—i.e., energy in the process of transfer from one body to another. After it has been transferred, energy is always designated according to its nature. Hence, heat transferred may become thermal energy, while work done may manifest itself in the form of mechanical energy.

All forms of energy are associated with motion. For example, any given body has kinetic energy if it is in motion. A tensioned device such as a bow or spring, though at rest, has the potential for creating motion; it contains potential energy because of its configuration. Similarly, nuclear energy is potential energy because it results from the configuration of subatomic particles in the nucleus of an atom.

Energy can be neither created nor destroyed but only changed from one form to another. This principle is known as the conservation of energy or the first law of thermodynamics. For example, when a box slides down a hill, the potential energy that the box has from being located high up on the slope is converted to kinetic energy, energy of motion. As the box slows to a stop through friction, the kinetic energy from the box's motion is converted to thermal energy that heats the box and the slope.

Energy can be converted from one form to another in various other ways. Usable mechanical or electrical energy is, for instance, produced by many kinds of devices, including fuel-burning heat engines, generators, batteries, fuel cells, and magnetohydrodynamic systems.

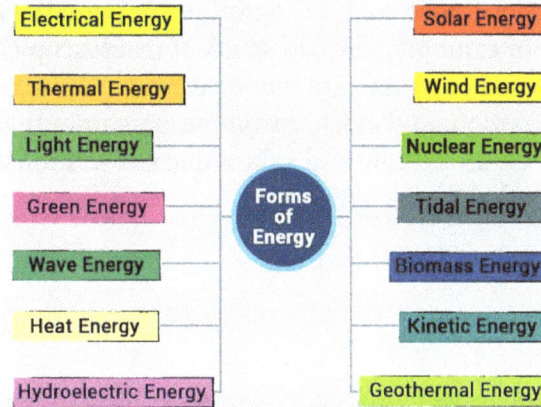

There are two types of energy:

1. Kinetic energy (working)

2. Potential energy (static)

There are two forms of energy sources:

1. Renewable source of energy

2. Non-renewable source of energy

Following are the examples of types of energy based on their sources:

1. Renewable source:

    a. Solar energy

    b. Wind energy

    c. Geothermal energy

2. Non-renewable source:

    a. Natural gas

    b. Coal

    c. Petroleum products

Unit of Energy

The SI unit of energy is Joules (J) which is nothing but a term for Newton-meter. When a certain amount of force (Newton) is applied to an object and it moved a certain distance (meters), then the energy applied is said to be Joules (newton-meters).

Types of Energy

There are different forms of energy but the distinction between them is not always clear. As Richard Feynman, a famous physicist once said, "The notions of potential and kinetic energy depend on a notion of length scale. For example, one can speak of macroscopic potential and kinetic energy, which do not include thermal potential and kinetic energy. Also what is called chemical potential energy is a macroscopic notion, and closer examination shows that it is really the sum of the potential and kinetic energy on the atomic and subatomic scale. Similar remarks apply to nuclear "potential" energy and most other forms of energy."

Some important types of energy and their features –

Kinetic Energy

The energy in motion is known as Kinetic Energy. For example a moving ball, flowing water, etc.

$$\text{Kinetic Energy} = \frac{1}{2} m \times v^2$$

Where,

      m = Mass of the object

      v = Velocity of the object

Potential Energy

This is the energy stored in an object and is measured by the amount of work done. For example, a pen on a table, water in a lake, etc.

$$\text{Potential Energy} = m \times g \times h$$

Where,

      m = Mass of the object (in kilograms)

      g = Acceleration due to gravity

      h = Height in meters

Mechanical Energy

It is the sum of potential energy and kinetic energy that is the energy associated with the motion &

position of an object is known as Mechanical energy. Thus, we can derive the formula of mechanical energy as –

Mechanical Energy = Kinetic Energy + Potential Energy

$$\text{Mechanical Energy} = \frac{1}{2} m \times v^2 + m \times g \times h$$

Solar Energy

The light and heat from the sun, harnessed using technologies like, solar heating, photovoltaics, solar thermal energy, solar architecture, and artificial photosynthesis is known as solar energy. It is the prime source of renewable energy.

Wind Energy

It is one of the various forms of energy. The energy present in the flow of wind, used by wind turbines is called wind energy. This energy is a major cheap source to produce electricity. In this phenomena, the kinetic energy of the wind is converted into mechanical power.

Nuclear Energy

The energy present in the nucleus of an atom is known as nuclear energy. The particles of an atom are tiny and need the energy to hold themselves. Nuclear energy is that enormous energy in the bonds of an atom which helps to hold the atom together. Nuclear energy can be used to make electricity.

Geothermal Energy

The energy or heat present inside the Earth is known as geothermal energy. It is a cheap & convenient heat and power resource and use of this energy don't have a side effect like greenhouse gas emission etc.

Tidal Energy

Tidal energy or tidal power is a form of hydropower (energy present in water), which converts the energy present in the tides to produce electricity.

Biomass Energy

Biomass is organic matter obtained from living organisms. The energy produced from biomass is called biomass energy.

Electrical Energy

The energy caused by moving electric charges is known as electrical energy. Electric energy is a type of kinetic energy as the electrical charges moves.

Thermal Energy

Thermal energy is the energy obtained from heat. It is a microscopic, disordered equivalent of mechanical energy.

There may be instances where an object posses more than one type of energy. For example, boiling water, posses both kinetic and potential energy along with heat energy.

Law of Conservation of Energy

The law of conservation of energy is one of the basic laws in physics. It governs the microscopic motion of individual atoms in a chemical reaction. The law of conservation of energy states that "In a closed system, i.e., a system that isolated from its surroundings, the total energy of the system is conserved." According to the law, the total energy in a system is conserved even though the transformation of energy occurs. Energy can neither be created nor destroyed, it can only be converted from one form to another.

Types of Energy Examples

Following are the examples of types of energy:

- Kinetic energy: A child swinging on a swing with no negative value irrespective of the to and fro motion.

- Gravitational energy: The atmosphere of the earth is held due to gravitational energy.

- Chemical energy: Energy stored in an electrochemical cell.

The understanding and development of energy are crucial for societal development. Our ability to utilize energy effectively improves the quality of life. It is hard to imagine life without energy.

Momentum and Collision

In classical mechanics, the momentum (SI unit kg m/s) of an object is the product of the mass and velocity of the object. Conceptually, the momentum of a moving object can be thought of as how difficult it would be to stop the object. As such, it is a natural consequence of Newton's first and second laws of motion. Having a lower speed or having less mass (how we measure inertia) results in having less momentum.

Momentum is a conserved quantity, meaning that the total momentum of any closed system (one not affected by external forces, and whose internal forces are not dissipative as heat or light) cannot be changed.

The concept of momentum in classical mechanics was originated by a number of great thinkers and experimentalists. René Descartes referred to mass times velocity as the fundamental force of motion. Galileo in his Two New Sciences used the term "impeto" (Italian), while Newton's

Laws of Motion uses motus (Latin), which has been interpreted by subsequent scholars to mean momentum.

Momentum in Newtonian Mechanics

If an object is moving in any reference frame, then it has momentum in that frame. It is important to note that momentum is frame dependent. That is, the same object may have a certain momentum in one frame of reference, but a different amount in another frame. For example, a moving object has momentum in a reference frame fixed to a spot on the ground, while at the same time having zero momentum in a reference frame that is moving along with the object.

The amount of momentum that an object has depends on two physical quantities—the mass and the velocity of the moving object in the frame of reference. In physics, the symbol for momentum is usually denoted by a small bold p (bold because it is a vector); so this can be written:

$$p = mv$$

where:

p is the momentum

m is the mass

v the velocity

The origin of the use of p for momentum is unclear.

The velocity of an object at a particular instant is given by its speed and the direction of its motion at that instant. Because momentum depends on and includes the physical quantity of velocity, it too has a magnitude and a direction and is a vector quantity. For example, the momentum of a five-kg bowling ball would have to be described by the statement that it was moving westward at two m/s. It is insufficient to say that the ball has ten kg m/s of momentum because momentum is not fully described unless its direction is also given.

Momentum for a System

Relating to Mass and Velocity

The momentum of a system of objects is the vector sum of the momenta of all the individual objects in the system.

$$\mathbf{p} = \sum_{i=1}^{n} m_i \vec{v}_i = m_1 \vec{v}_1 + m_2 \vec{v}_2 + m_3 \vec{v}_3 + \ldots + m_n \vec{v}_n$$

where,

p is the momentum

m_i is the mass of object i

$\vec{v_i}$ the vector velocity of object $_i$

n is the number of objects in the system

Relating to Force

Force is equal to the rate of change of momentum:

$$\mathbf{F} = \frac{d\mathbf{p}}{dt}$$

In the case of constant mass and velocities much less than the speed of light, this definition results in the equation $\mathbf{F} = m\mathbf{a}$ — commonly known as Newton's second law.

If a system is in equilibrium, then the change in momentum with respect to time is equal to zero:

$$\mathbf{F} = \frac{d\mathbf{p}}{dt} = m\mathbf{a} = 0$$

Conservation of Momentum

The principle of conservation of momentum states that the total momentum of a closed system of objects (which has no interactions with external agents) is constant. One of the consequences of this is that the center of mass of any system of objects will always continue with the same velocity unless acted on by a force outside the system.

In an isolated system (one where external forces are absent) the total momentum will be constant—this is implied by Newton's first law of motion. Newton's third law of motion, the law of reciprocal actions, which dictates that the forces acting between systems are equal in magnitude, but opposite in sign, is due to the conservation of momentum.

Since momentum is a vector quantity it has direction. Thus, when a gun is fired, although overall movement has increased compared to before the shot was fired, the momentum of the bullet in one direction is equal in magnitude, but opposite in sign, to the momentum of the gun in the other direction. These then sum to zero which is equal to the zero momentum that was present before either the gun or the bullet was moving.

Collisions

Momentum has the special property that, in a closed system, it is always conserved, even in collisions. Kinetic energy, on the other hand, is not conserved in collisions if they are inelastic (where two objects collide and move off together at the same velocity). Since momentum is conserved it can be used to calculate unknown velocities following a collision.

A common problem in physics that requires the use of this fact is the collision of two particles. Since momentum is always conserved, the sum of the momenta before the collision must equal the sum of the momenta after the collision:

$$m_1\mathbf{u}_1 + m_2\mathbf{u}_2 = m_1\mathbf{v}_1 + m_2\mathbf{v}_2$$

where:

u signifies vector velocity before the collision

v signifies vector velocity after the collision.

Usually, we either only know the velocities before or after a collision and would like to also find out the opposite. Correctly solving this problem means you have to know what kind of collision took place. There are two basic kinds of collisions, both of which conserve momentum:

- Elastic collisions conserve kinetic energy as well as total momentum before and after collision.

- Inelastic collisions don't conserve kinetic energy, but total momentum before and after collision is conserved.

Elastic Collisions

A collision between two pool balls is a good example of an almost totally elastic collision. In addition to momentum being conserved when the two balls collide, the sum of kinetic energy before a collision must equal the sum of kinetic energy after:

$$\frac{1}{2}m_1v_{1,i}^2 + \frac{1}{2}m_2v_{2,i}^2 = \frac{1}{2}m_1v_{1,f}^2 + \frac{1}{2}m_2v_{2,f}^2$$

Since the one-half factor is common to all the terms, it can be taken out right away.

Head-on Collision (1 Dimensional)

In the case of two objects colliding head on we find that the final velocity,

$$v_{1,f} = \left(\frac{m_1 - m_2}{m_1 + m_2}\right)v_{1,i} + \left(\frac{2m_2}{m_1 + m_2}\right)v_{2,i}$$

$$v_{2,f} = \left(\frac{2m_1}{m_1 + m_2}\right)v_{1,i} + \left(\frac{m_2 - m_1}{m_1 + m_2}\right)v_{2,i}$$

which can then easily be rearranged to,

$$m_{1,f} \cdot v_{1,f} + m_{2,f} \cdot v_{2,f} = m_{1,i} \cdot v_{1,i} + m_{2,i} \cdot v_{2,i}$$

Special Case: m1 much Greater than m2

Now consider if [[mass]] of one body say m1 is far more than m2 (m1>>m2). In that case m1+m2 is approximately equal to m1. And m1-m2 is approximately equal to m1.

Put these values in the above equation to calculate the value of v2 after collision. The expression changes to v2 final is 2*v1-v2. Its physical interpretation is in case of collision between two body one of which is very heavy, the lighter body moves with twice the velocity of the heavier body less its actual velocity but in opposite direction.

Special Case: m1 Equal to m2

Another special case is when the collision is between two bodies of equal mass. Say body m1 moving at velocity v1 strikes body m2 that is at rest (v2). Putting this case in the equation derived above we will see that after the collision, the body that was moving (m1) will start moving with velocity v2 and the mass m2 will start moving with velocity v1. So there will be an exchange of velocities.

Now suppose one of the masses, say m2, was at rest. In that case after the collision the moving body, m1, will come to rest and the body that was at rest, m2, will start moving with the velocity that m1 had before the collision.

Please note that all of these observations are for an elastic collision.

This phenomenon called "Newton's cradle," one of the most well-known examples of conservation of momentum, is a real life example of this special case.

Multi-dimensional Collisions

In the case of objects colliding in more than one dimension, as in oblique collisions, the velocity is resolved into orthogonal components with one component perpendicular to the plane of collision and the other component or components in the plane of collision. The velocity components in the plane of collision remain unchanged, while the velocity perpendicular to the plane of collision is calculated in the same way as the one-dimensional case.

For example, in a two-dimensional collision, the momenta can be resolved into x and y components. We can then calculate each component separately, and combine them to produce a vector result. The magnitude of this vector is the final momentum of the isolated system.

Inelastic Collisions

A common example of a perfectly inelastic collision is when two snowballs collide and then stick together afterwards. This equation describes the conservation of momentum:

$$m_1 \mathbf{v}_{1,i} + m_2 \mathbf{v}_{2,i} = \left(m_1 + m_2 \right) \mathbf{v}_f$$

It can be shown that a perfectly inelastic collision is one in which the maximum amount of kinetic energy is converted into other forms. For instance, if both objects stick together after the collision and move with a final common velocity, one can always find a reference frame in which the objects are brought to rest by the collision and 100 percent of the kinetic energy is converted.

Reference

- Classical-mechanics: chegg.com, Retrieved 2 June, 2019

- Basics-of-kinematics, boundless-physics: lumenlearning.com, Retrieved 9 January, 2019

- Dynamics: real-world-physics-problems.com, Retrieved 19 March, 2019

- The-Meaning-of-Force: physicsclassroom.com, Retrieved 21 July, 2019

- What-are-newtons-laws-of-motion: thoughtco.com, Retrieved 18 April, 2019

- Newton's-laws-of-motion, force-and-laws-of-motion, physics: toppr.com, Retrieved 28 August, 2019

- What-are-newtons-laws-of-motion: thoughtco.com, Retrieved 13 July, 2019

- Circular-motion, mechanics, physics: physicsnet.co.uk, Retrieved 17 January, 2019

- Rotational-motion: byjus.com, Retrieved 7 May, 2019

- Energy, science: britannica.com, Retrieved 12 August, 2019

- Energy, physics: byjus.com, Retrieved 15 June, 2019

- Momentum: newworldencyclopedia.org, Retrieved 26 February, 2019

3

Understanding Gravity

Gravity is a force by which all things with mass and energy including planets, starts, and galaxies are brought toward one and another. A potential energy that a physical object with mass has due to gravity in relation to another massive object is known as gravitational potential energy. This chapter discusses in detail the theories and methodologies related to gravity.

Gravity, also called gravitation is the universal force of attraction acting between all matter. It is by far the weakest known force in nature and thus plays no role in determining the internal properties of everyday matter. On the other hand, through its long reach and universal action, it controls the trajectories of bodies in the solar system and elsewhere in the universe and the structures and evolution of stars, galaxies, and the whole cosmos. On Earth all bodies have a weight, or downward force of gravity, proportional to their mass, which Earth's mass exerts on them. Gravity is measured by the acceleration that it gives to freely falling objects. At Earth's surface the acceleration of gravity is about 9.8 metres (32 feet) per second per second. Thus, for every second an object is in free fall, its speed increases by about 9.8 metres per second. At the surface of the Moon the acceleration of a freely falling body is about 1.6 metres per second per second.

The works of Isaac Newton and Albert Einstein dominate the development of gravitational theory. Newton's classical theory of gravitational force held sway from his Principia, published in 1687, until Einstein's work in the early 20th century. Newton's theory is sufficient even today for all but the most precise applications. Einstein's theory of general relativity predicts only minute quantitative differences from the Newtonian theory except in a few special cases. The major significance of Einstein's theory is its radical conceptual departure from classical theory and its implications for further growth in physical thought.

The launch of space vehicles and developments of research from them have led to great improvements in measurements of gravity around Earth, other planets, and the Moon and in experiments on the nature of gravitation.

Newton's Law of Gravitation

The attractive force between two particles:

$$F = G \frac{m_1 m_2}{r^2}$$

where $G = 6.67 \times 10^{-11} \ N.m^2 / kg^2$ is the universal gravitational constant.

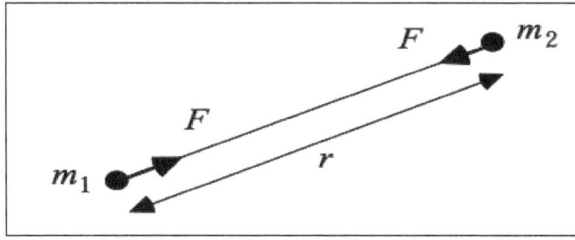

- Particle 1 feels a pull toward particle 2 and particle 2 feels a pull towards particle 1 -- action-reaction forces.

- The law is for pairs of "point-like" particles.

- Every particle in the universe pulls on every other particle in the universe, e.g. the moon is pulling on you now.

- The force does not depend on what is between two objects, i.e. it cannot be shielded by a material (e.g. wall) between them.

- This is one of the four fundamental forces.

Principle of Superposition

The total force on a point particle is equal to the sum of all the forces on the particle.

$$\vec{F}_1 = \vec{F}_{12} + \vec{F}_{13} + \vec{F}_{14} + \vec{F}_{1n}$$

$$= \sum_{i=2}^{n} \vec{F}_{1i}$$

- $\vec{F}_{1i}$: force of the ith particle on particle 1.

For a real object with a continuous distribution of particles:

$$dF_1 = G \frac{m_1 dm}{r^2}$$

$$F_1 = \int dF_1$$

$$= \int G \frac{m_1}{r^2} dm$$

Gravitational Force from a Thin Ring

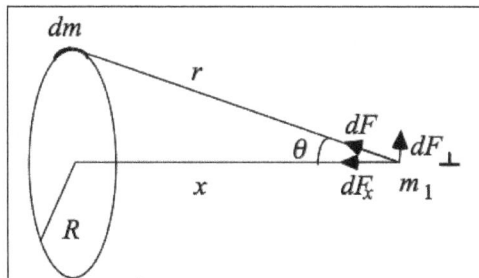

$$F_x = \int dF_x$$

$$= \int G \frac{m_1 dm}{r^2} \cos q$$

$$= \int G \frac{m_1 dm}{r^2} \frac{x}{r}$$

$$= \int G \frac{m_1 x dm}{r^3}$$

$$= \int G \frac{m_1 x dm}{\left(x^2 + R^2\right)^{\frac{3}{2}}}$$

$$= G \frac{m_1 m x}{\left(x^2 + R^2\right)^{\frac{3}{2}}}$$

The force points toward the ring.

$$F_x = 0 \text{ if } x = 0$$

The force perpendicular to x ($dF_\perp$) cancels by symmetry. Use symmetry to simplify your problem.

Other Newton's results:

- A uniform spherical shell of matter attracts a particle outside as if all the shell mass was concentrated at the center.

- Similarly for a sphere of matter.

- Like the case of a ring, a particle inside a spherical shell of matter feels zero gravitational force from the shell.

Acceleration due to Gravity:

Force on a mass m on Earth's surface:

$$F_g = \frac{GmM_{Earth}}{R_{Earth}^2}$$

$$= m \frac{GM_{Earth}}{R_{Earth}^2}$$

$$= \frac{GM_{Earth}}{R_{Earth}^2} = \frac{\left(6.67 \times 10^{-11} m^3 / kg / s^2\right)\left(5.98 \times 10^{24} kg\right)}{\left(6.37 \times 10^6 m\right)^2}$$

$$= 9.83 \quad m/s^2$$

$$= g$$

For any planet, the acceleration of gravity at its surface is:

$$g_{planet} = \frac{GmM_{planet}}{R_{planet}^2}$$

Example:

What is the acceleration due to gravity for (a) an airplane flying at an altitude of 10 km, (b) a shuttle at 300 km, (c) a geosynchronous satellite at 36000 km?

(a)

$$g_{plane} = \frac{GM_{Earth}}{R_{Earth} + h_{plane}}$$

$$= \frac{\left(6.67 \times 10^{-11} m^3 / kg / s^2\right)\left(5.98 \times 10^{24} kg\right)}{\left(6.37 \times 10^6 m + 1 \times 10^4 \ m\right)^2}$$

$$= 9.79 \ \ m/s^2$$

$$= g$$

(b)

$$g_{shuttle} = \frac{GM_{Earth}}{\left(R_{Earth} + h_{plane}\right)}$$

$$= \frac{\left(6.67 \times 10^{-11} m^3 / kg / s^2\right)\left(5.98 \times 10^{24} kg\right)}{\left(6.37 \times 10^6 m + 3 \times 10^5 \ m\right)^2}$$

$$= 9.00 \ \ m/s^2$$

Little different from g on Earth. Astronauts floating in a shuttle is not due to zero gravity.

(c)

$$g_{satelite} = \frac{GM_{Earth}}{\left(R_{Earth} + h_{shuttle}\right)^2}$$

$$= \frac{\left(6.67 \times 10^{-11} m^3 / kg / s^2\right)\left(5.98 \times 10^{24} kg\right)}{\left(6.37 \times 10^6 m + 3.6 \times 10^7 \ m\right)^2}$$

$$= 0.22 \ \ m/s^2$$

Much lower than g on Earth but gravity still pulls on objects in space.

Example:

For a spaceship between the Earth and moon, at what distance from the Earth will the net gravitational force be zero?

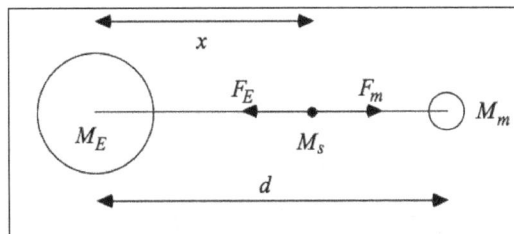

$$\sum F_x = F_{Earth} - F_{moon} = 0$$

$$\frac{GM_E M_s}{x^2} - \frac{GM_m M_s}{(x-d)^2} = 0$$

$$\frac{M_E}{x^2} - \frac{M_m}{(x-d)^2} = 0$$

$$(x-d)^2 - \frac{M_m}{M_E} x^2 = 0$$

$$x^2 - 2xd + d^2 - rx^2 = 0$$

$$(1-r)x^2 - 2xd + d^2 = 0$$

$$x = \frac{2d \pm \sqrt{4d^2 - 4(1-r)d^2}}{2(1-r)}$$

$$= \frac{2d \pm \sqrt{4rd^2}}{2(1-r)}$$

$$= \frac{2d \pm 2\sqrt{rd}}{2(1-r)}$$

$$= \frac{1 \pm \sqrt{r}}{1-r} d$$

- $x = \frac{1+\sqrt{r}}{1-r} d > d \Rightarrow$ unphysical

- $x = \frac{1-\sqrt{r}}{1-r} d < d \Rightarrow$ OK

Example:

Calculate the gravitational force due to a hollowed sphere, assuming that the mass of the sphere was M before hollowing.

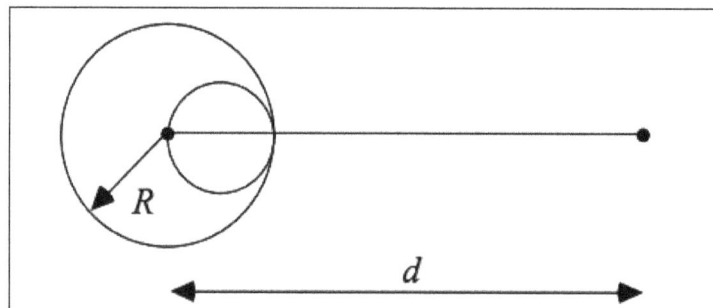

Calculating the force by integrating over the solid part of the sphere is difficult.

Use Principle of Superposition

F = F (solid sphere) − F(hollowed material)

$$= \frac{GMm}{d^2} - \frac{GM_{hm}m}{\left(d - \dfrac{R}{2}\right)^2}$$

$$\frac{GMm}{d^2} - \frac{GM_{hm}m}{\left(d - \dfrac{R}{2}\right)^2} \times M \times \frac{\dfrac{4}{3}\pi\left(\dfrac{R}{2}\right)^3}{\dfrac{4}{3}\pi R^3}$$

$$= \frac{GMm}{d^2} - \frac{GM_{hm}m}{\left(d - \dfrac{R}{2}\right)^2} \times \frac{1}{8}$$

$$\frac{GMm}{d^2}\left[1 - \frac{1}{2\left(2 - \dfrac{R}{D}\right)^2}\right]$$

Satellite

Gravitational Force is the centripetal force that keeps a satellite moving in a circular orbit.

$$Fc = Fg$$

$$m\frac{v^2}{r} = \frac{GM_E m}{r^2}$$

$$v^2 = \frac{GM_E}{r}$$

$$v = \sqrt{\frac{GM_E}{r}}$$

The speed is independent of the mass of the satellite, a penny and a semi-truck would orbit at the same speed for the same radius.

- A satellite orbiting further away from the Earth has lower speed.

- The period of the orbit:

$$Tc = \frac{2pr}{v} = \frac{2pr}{\sqrt{GME/r}} = \sqrt{\frac{4p^2 r^3}{GM_E}}$$

- Communication satellites are positioned in orbit so that they "appear" stationary in the sky by placing them in a circular orbit with a period of 1 day.

What is the radius of the orbit of a geosynchronous satellite?

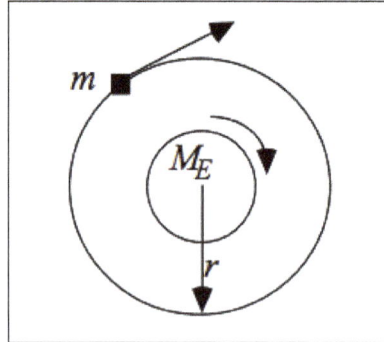

$$r^3 = \frac{GM_E T_c^2}{4\pi^2}$$

$$= \frac{(6.67 \times 10^{-11} m^3 kg^{-1} S^{-2})(5.98 \times 10^{24} kg)(86400s)^2}{4\pi^2}$$

$$r = 4.2 \times 10^7 \, m$$
$$= 42,000 km$$

Apparent Weightlessness

The weight of an object is defined as the force of gravity that the object feels. For an astronaut in a spacecraft, the weight is just:

$$F_g = \frac{GMm}{r^2}$$

where M is the mass of the earth and m is the mass of the astronaut. However, we see pictures of astronauts floating around in the space shuttle in an apparently weightless environment. Are they actually weightless?

Since we define weight to be the force of gravity, the astronauts are not weightless. We can define an apparent weight as being the normal force between an object and the support (e.g. floor).

Consider a ride in elevator:

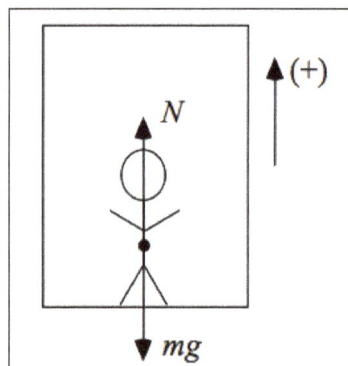

If the elevator is accelerating upward,

$$N - mg = ma$$
$$N = mg + ma$$
$$> mg$$

The reading on a scale will give an apparent weight larger than the actual weight because the scale measures the normal force.

- If the elevator is accelerating downward, the apparent weight is smaller than the actual weight because normal force is less than mg.

- If the elevator is in a free fall,

$$N - mg = -mg$$
$$N = 0$$

The reading on a scale would be zero, giving a zero apparent weight.

Consider an astronaut in a shuttle:

Shuttle:

$$G \frac{M_E m_s}{r^2} + N = m_s \frac{v^2}{r}$$

$$\frac{N}{m_s} = \frac{v^2}{r} - G \frac{M_E}{r^2}$$

Astronaut:

$$G \frac{M_E m_a}{r^2} - N = m_a \frac{v^2}{r}$$

$$\frac{N}{m_a} = G \frac{M_E}{r^2} - \frac{v^2}{r}$$

$$\frac{N}{m_s} + \frac{N}{m_a} = 0$$

$$N = 0$$

Astronauts floating in a shuttle. However, astronauts are not weightless:

$$F_g = G \ \frac{M_E m_a}{r^2} \ \neq \ 0$$

Kepler's Laws

Kepler's Law states that the planets move around the sun in elliptical orbits with the sun at one focus. There are three different Kepler's Laws. Law of Orbits, Areas, and Periods.

Kepler's three law:

1. Kepler's Law of Orbits – The Planets move around the sun in elliptical orbits with the sun at one focus.

2. Kepler's Law of Areas – The line joining a planet to the Sun sweeps out equal areas in equal interval of time.

3. Kepler's Law of Periods – The square of the time period of the planet is directly proportional to the cube of the semimajor axis of its orbit.

Kepler's 1st Law of Orbits

This law is popularly known as the law of orbits. The orbit of any planet is an ellipse around the Sun with Sun at one of the two foci of an ellipse. We know that planets revolve around the Sun in a circular orbit. But according to Kepler, he said that it is true that planets revolve around the Sun, but not in a circular orbit but it revolves around an ellipse. In an ellipse, we have two focus. Sun is located at one of the foci of the ellipse.

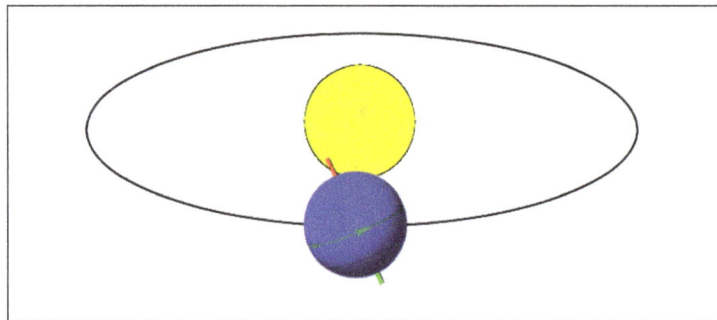

Kepler's 2nd Law of Areas

This law is known as the law of areas. The line joining a planet to the Sun sweeps out equal areas in equal interval of time. The rate of change of area with time will be constant. We can see in the above figure, the Sun is located at the focus and the planets revolve around the Sun.

Assume that the planet starts revolving from point P_1 and travels to P_2 in a clockwise direction. So it revolves from point P_1 to P_2, as it moves the area swept from P_1 to P_2 is Δt. Now the planet moves future from P_3 to P_4.and the area covered is Δt.

As the area traveled by the planet from P_1 to P_2 and P_3 to P_4 is equal, therefore this law is known as the Law of Area. That is the aerial velocity of the planets remains constant. When a planet is nearer to the Sun it moves fastest as compared to the planet far away from the Sun.

Kepler's 3rd Law of Periods:

This law is known as the law of Periods. The square of the time period of the planet is directly proportional to the cube of the semimajor axis of its orbit.

$$T^2 \propto a^3$$

That means the time 't' is directly proportional to the cube of the major axis. Let us derive the equation of Kepler's 3rd law. Let us suppose,

- m = mass of the planet

- M = mass of the Sun

- v = velocity in the orbit

So, there has to be a force of gravitation between the Sun and the planet.

$$F = \frac{GmM}{r^2}$$

Since it is moving in an elliptical orbit, there has to be a centripetal force.

$$F_c = \frac{mv^2}{r^2}$$

$$Now,\ F = F_c$$

$$\Rightarrow \frac{GM}{r} = v^2$$

$$Also,\ v = \frac{circumference}{time} = \frac{2\pi r}{t}$$

Combining the above equations, we get

$$\Rightarrow \frac{GM}{r} = \frac{4\pi^2 r^2}{T^2}$$

$$T^2 = \frac{4\pi^2 r^3}{GM}$$

$$\Rightarrow T^2 \propto r^3$$

Gravitational Potential Energy

The gravitational energy or gravitational potential energy is the energy stored in the object because of its height above the ground within the gravitational field. We are all experience the force of gravity within the gravitational field of the earth.

If the work is done on the object against the gravitational force, the object will gains gravitational potential energy.

For example, when an object is lifted up to a height 'h', work is done against the gravitational force which is pulling it down. This work done on the object is stored as potential energy. If this lifted object is released from height 'h' then its potential energy gets converted into the energy of motion or kinetic energy.

The amount of gravitational potential energy an object has depends on mass of the object and height of the object above ground. Objects that are at large height above the ground have more potential energy. Similarly, objects that are at small height above the ground have less potential energy.

Objects that have more mass have more potential energy. Similarly, objects that have less mass have less potential energy.

Let's take for example, object C has more potential energy than object B because object C has more mass than object B. Similarly, object A has less potential energy than objects C because object A is present at less height.

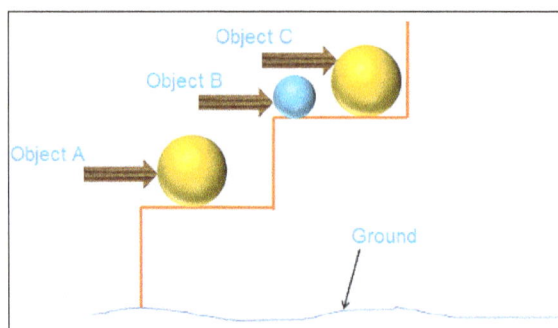

Gravitational energy

The gravitational potential energy is given as,

$$PE_{grav} = m \times g \times h$$

Where:

PE_{grav} = potential energy

m = mass of the object

g = acceleration due to gravity = constant

h = height of the object above the earth surface

Reference

- Gravity-physics, science: britannica.com, Retrieved 5 March, 2019

- Keplers-law, gravitation, physics: toppr.com, Retrieved 18 February, 2019

- Gravitational-energy, potential-energy, energy, physics: physics-and-radio-electronics.com, Retrieved 14 May, 2019

4

Thermodynamics and Statistical Mechanics

Thermodynamics is a domain of physics that is concerned with heat and temperature and their relation to radiation, energy, work and properties of matter. Statistical mechanics is a subset that deals with the study of any physical system that has a large number of degrees of freedom. The chapter closely examines the key concepts of thermodynamics and statistical mechanics to provide an extensive understanding of the subject.

Heat

In thermodynamics, heat has a very specific meaning that is different from how we might use the word in everyday speech. Scientists define heat as thermal energy transferred between two systems at different temperatures that come in contact. Heat is written with the symbol q or Q, and it has units of Joules (J).

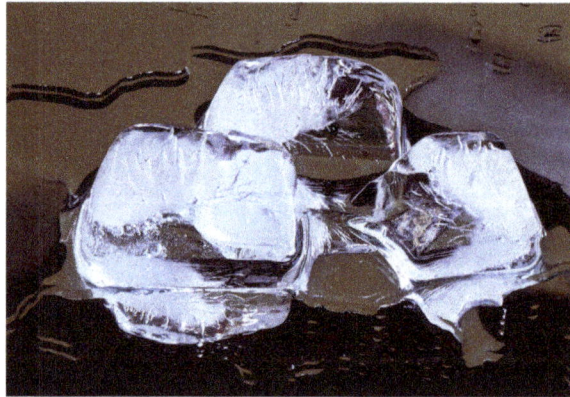

Heat is transferred from the surroundings to the ice, causing the phase change from ice to water.

Heat is sometimes called a process quantity, because it is defined in the context of a process by which energy can be transferred. We don't talk about a cup of coffee containing heat, but we can talk about the heat transferred from the cup of hot coffee to your hand. Heat is also an extensive property, so the change in temperature resulting from heat transferred to a system depends on how many molecules are in the system.

Relationship between Heat and Temperature

Heat and temperature are two different but closely related concepts. Note that they have different units: temperature typically has units of degrees Celsius (°C) or Kelvin (K), and heat has units of energy, Joules (J). Temperature is a measure of the average kinetic energy of the atoms or molecules in the system. The water molecules in a cup of hot coffee have a higher average kinetic energy than the water molecules in a cup of iced tea, which also means they are moving at a higher velocity. Temperature is also an intensive property, which means that the temperature doesn't change no matter how much of a substance you have (as long as it is all at the same temperature!). This is why chemists can use the melting point to help identify a pure substance – the temperature at which it melts is a property of the substance with no dependence on the mass of a sample.

On an atomic level, the molecules in each object are constantly in motion and colliding with each other. Every time molecules collide, kinetic energy can be transferred. When the two systems are in contact, heat will be transferred through molecular collisions from the hotter system to the cooler system. The thermal energy will flow in that direction until the two objects are at the same temperature. When the two systems in contact are at the same temperature, we say they are in thermal equilibrium.

Heat Capacity: Converting between Heat and Change in Temperature

How can we measure heat? Here are some things we know about heat so far:

- When a system absorbs or loses heat, the average kinetic energy of the molecules will change. Thus, heat transfer results in a change in the system's temperature as long as the system is not undergoing a phase change.

- The change in temperature resulting from heat transferred to or from a system depends on how many molecules are in the system.

We can use a thermometer to measure the change in a system's temperature. How can we use the change in temperature to calculate the heat transferred?

In order to figure out how the heat transferred to a system will change the temperature of the system, we need to know at least 2 things:

- The number of molecules in the system

- The heat capacity of the system

The heat capacity tells us how much energy is needed to change the temperature of a given substance assuming that no phase changes are occurring. There are two main ways that heat capacity is reported. The specific heat capacity (also called specific heat), represented by the symbol c or C, is how much energy is needed to increase the temperature of one gram of a substance by 1 °C or 1 K. Specific heat capacity usually has units of $\frac{J}{grams \cdot K}$. The molar heat capacity, C_m C_{mol}, measures the amount of thermal energy it takes to raise the temperature of one mole of a substance by 1 °C or 1 K and it usually has units of $\frac{J}{mol \cdot K}$. For example, the heat capacity of lead might be given as the specific heat capacity, 0.129 $\frac{J}{g \cdot K}$, or the molar heat capacity, 26.65 $\frac{J}{mol \cdot K}$.

Calculating q Using the Heat Capacity

We can use the heat capacity to determine the heat released or absorbed by a material using the following formula:

$$q = m \times C \times \Delta T$$

where m is the mass of the substance (in grams), C is the specific heat capacity, and ΔT, T is the change in temperature during the heat transfer. Note that both mass and specific heat capacity can only have positive values, so the sign of q will depend on the sign of ΔT, T. We can calculate ΔT, T using the following equation:

$$\Delta T = T_{final} - T_{initial}$$

Where T_{final} and T_{intial} can have units of either °C degree, K. Based on this equation, if q is positive (energy of the system increases), then our system increases in temperature and $T_{final} > T_{intial}$ is negative (energy of the system decreases), then our system's temperature decreases and $T_{final} < T_{intial}$.

Cooling a Cup of Tea

Let's say that we have 250mL, space, m, L of hot tea which we would like to cool down before we try to drink it. The tea is currently at 370K, and we'd like to cool it down to 350 K. How much thermal energy has to be transferred from the tea to the surroundings to cool the tea?

The hot tea will transfer heat to the surroundings as it cools.

We are going to assume that the tea is mostly water, so we can use the density and heat capacity of water in our calculations.

The specific heat capacity of water is $4.18 \dfrac{J}{g \cdot K}$, and the density of water is $1.00 \dfrac{g}{mL}$. We can calculate the energy transferred in the process of cooling the tea using the following steps:

1. Calculate the mass of the substance

We can calculate the mass of the tea/water using the volume and density of water:

$$m = 250 \; \cancel{mL} \times 1.00 \dfrac{g}{\cancel{mL}} = 250 \, g$$

2. Calculate the change in temperature, ΔT

We can calculate the change in temperature, ΔT, from the initial and final temperatures:

$$\Delta T = T_{final} - T_{initial}$$
$$= 350K - 370K$$
$$= -20K$$

Since the temperature of the tea is decreasing and ΔT is negative, we would expect q to also be negative since our system is losing thermal energy.

3. Solve for q

Now we can solve for the heat transferred from the hot tea using the equation for heat:

$$q = m \times C \times \Delta T$$

$$= 250 \; \cancel{g} \times 4.18 \frac{J}{g \cdot K} \times -20 \; \cancel{K}$$

Thus, we calculated that the tea will transfer 21000J space J of energy to the surroundings when it cools down from 370K to 350K.

Thermodynamics

Thermodynamics is a branch of physics that studies the effects of changes in temperature, pressure, and volume on physical systems at the macroscopic scale by analyzing the collective motion of their particles using statistics. In this context, heat means "energy in transit" and dynamics relates to "movement;" thus, thermodynamics is the study of the movement of energy and how energy instills movement. Historically, thermodynamics developed out of need to increase the efficiency of early steam engines.

The starting point for most thermodynamic considerations are the laws of thermodynamics, which postulate that energy can be exchanged between physical systems as heat or work. The first law of thermodynamics states a universal principle that processes or changes in the real world involve energy, and within a closed system the total amount of that energy does not change, only its form (such as from heat of combustion to mechanical work in an engine) may change. The second law gives a direction to that change by specifying that in any change in any closed system in the real world the degree of order of the system's matter and energy becomes less, or conversely stated, the amount of disorder (entropy) of the system increases.

In thermodynamics, interactions between large ensembles of objects are studied and categorized. Central to this are the concepts of system and surroundings. A system comprises particles whose average motions define the system's properties, which are related to one another through equations of state defining the relations between state variables such as temperature, pressure, volume, and entropy. State variables can be combined to express internal energy and thermodynamic potentials, which are useful for determining conditions for equilibrium and spontaneous processes.

With these tools, thermodynamics describes how systems respond to changes in their surroundings. This can be applied to a wide variety of topics in science and engineering, such as engines, phase transitions, chemical reactions, transport phenomena, and even black holes. The results of thermodynamics are essential for other fields of physics and for chemistry, chemical engineering, aerospace engineering, mechanical engineering, cell biology, biomedical engineering, and materials science to name a few.

Thermodynamics, with its insights into the relations between heat, energy, and work as exemplified in mechanical systems, provides a foundation for trying to understand the behavior and properties of biological, social, and economic systems, which generally maintain an ordered pattern only by consuming a sustained flow of energy.

Thermodynamic Systems

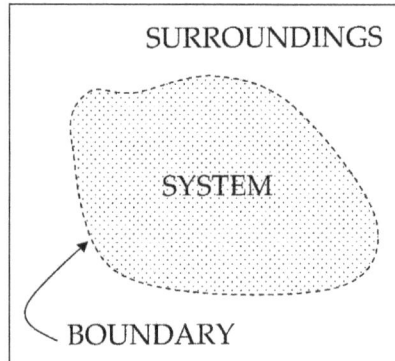

An important concept in thermodynamics is the "system." Everything in the universe except the system is known as surroundings. A system is the region of the universe under study. A system is separated from the remainder of the universe by a boundary which may or may not be imaginary, but which by convention delimits a finite volume. The possible exchanges of work, heat, or matter between the system and the surroundings take place across this boundary. Boundaries are of four types: Fixed, movable, real, and imaginary.

Basically, the "boundary" is simply an imaginary dotted line drawn around the volume of a something in which there is going to be a change in the internal energy of that something. Anything that passes across the boundary that effects a change in the internal energy of that something needs to be accounted for in the energy balance equation. That "something" can be the volumetric region surrounding a single atom resonating energy, such as Max Planck defined in 1900; it can be a body of steam or air in a steam engine, such as Sadi Carnot defined in 1824; it can be the body of a tropical cyclone, such as Kerry Emanuel theorized in 1986, in the field of atmospheric thermodynamics; it could also be just one nuclide (that is, a system of quarks) as some are theorizing presently in quantum thermodynamics.

For an engine, a fixed boundary means the piston is locked at its position; as such, a constant volume process occurs. In that same engine, a movable boundary allows the piston to move in and out. For closed systems, boundaries are real, while for open systems, boundaries are often imaginary. There are five dominant classes of systems:

1. Isolated Systems—matter and energy may not cross the boundary

2. Adiabatic Systems—heat must not cross the boundary

3. Diathermic Systems—heat may cross boundary

4. Closed Systems—matter may not cross the boundary

5. Open Systems—heat, work, and matter may cross the boundary (often called a control volume in this case)

As time passes in an isolated system, internal differences in the system tend to even out and pressures and temperatures tend to equalize, as do density differences. A system in which all equalizing processes have gone practically to completion is considered to be in a state of thermodynamic equilibrium.

In thermodynamic equilibrium, a system's properties are, by definition, unchanging in time. Systems in equilibrium are much simpler and easier to understand than systems that are not in equilibrium. Often, when analyzing a thermodynamic process, it can be assumed that each intermediate state in the process is at equilibrium. This will also considerably simplify the situation. Thermodynamic processes which develop so slowly as to allow each intermediate step to be an equilibrium state are said to be reversible processes.

Thermodynamic Parameters

The central concept of thermodynamics is that of energy, the ability to do work. As stipulated by the first law, the total energy of the system and its surroundings is conserved. It may be transferred into a body by heating, compression, or addition of matter, and extracted from a body either by cooling, expansion, or extraction of matter. For comparison, in mechanics, energy transfer results from a force which causes displacement, the product of the two being the amount of energy transferred. In a similar way, thermodynamic systems can be thought of as transferring energy as the result of a generalized force causing a generalized displacement, with the product of the two being the amount of energy transferred. These thermodynamic force-displacement pairs are known as conjugate variables. The most common conjugate thermodynamic variables are pressure-volume (mechanical parameters), temperature-entropy (thermal parameters), and chemical potential-particle number (material parameters).

Thermodynamic States

When a system is at equilibrium under a given set of conditions, it is said to be in a definite state. The state of the system can be described by a number of intensive variables and extensive variables. The properties of the system can be described by an equation of state which specifies the relationship between these variables. State may be thought of as the instantaneous quantitative description of a system with a set number of variables held constant.

Thermodynamic Processes

A thermodynamic process may be defined as the energetic change of a thermodynamic system proceeding from an initial state to a final state. Typically, each thermodynamic process is distinguished from other processes in energetic character, according to what parameters, such as temperature, pressure, or volume, etc., are held fixed. Furthermore, it is useful to group these processes into pairs, in which each variable held constant is one member of a conjugate pair. The seven most common thermodynamic processes are shown below:

1. An isobaric process occurs at constant pressure

2. An isochoric process, or isometric/isovolumetric process, occurs at constant volume

3. An isothermal process occurs at a constant temperature

4. An adiabatic process occurs without loss or gain of heat

5. An isentropic process (reversible adiabatic process) occurs at a constant entropy

6. An isenthalpic process occurs at a constant enthalpy. Also known as a throttling process or wire drawing

7. A steady state process occurs without a change in the internal energy of a system.

Thermodynamic Potentials

As can be derived from the energy balance equation on a thermodynamic system there exist energetic quantities called thermodynamic potentials, being the quantitative measure of the stored energy in the system. The five most well known potentials are:

Internal energy	U
Helmholtz free energy	$A = U - TS$
Enthalpy	$H = U + PV$
Gibbs free energy	$G = U + PV - TS$
Grand potential	$\Phi_G = U - TS - \mu N$

Potentials are used to measure energy changes in systems as they evolve from an initial state to a final state. The potential used depends on the constraints of the system, such as constant temperature or pressure. Internal energy is the internal energy of the system, enthalpy is the internal energy of the system plus the energy related to pressure-volume work, and Helmholtz and Gibbs energy are the energies available in a system to do useful work when the temperature and volume or the pressure and temperature are fixed, respectively.

Zeroth Law of Thermodynamics

The Zeroth Law of Thermodynamics states that systems in thermal equilibrium are at the same temperature. Systems are in thermal equilibrium if they do not transfer heat, even though they are in a position to do so, based on other factors. For example, food that's been in the refrigerator overnight is in thermal equilibrium with the air in the refrigerator: heat no longer flows from one source (the food) to the other source (the air) or back.

What the Zeroth Law of Thermodynamics means is that temperature is something worth measuring, because it indicates whether heat will move between objects. This will be true regardless of how the objects interact. Even if two objects don't touch, heat may still flow between them, such as by radiation (as from a heat lamp). However, according to the Zeroth Law of Thermodynamics, if the systems are in thermal equilibrium, no heat flow will take place.

There are more formal ways to state the Zeroth Law of Thermodynamics, which is commonly stated in the following manner:

Let A, B, and C be three systems. If A and C are in thermal equilibrium, and A and B are in thermal equilibrium, then B and C are in thermal equilibrium.

This statement is represented symbolically in. Temperature is not mentioned explicitly, but it's implied that temperature exists. Temperature is the quantity that is always the same for all systems in thermal equilibrium with one another.

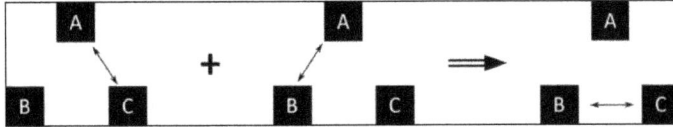

Zeroth Law of Thermodynamics: The double arrow represents thermal equilibrium between systems. If systems A and C are in equilibrium, and systems A and B are in equilibrium, then systems B and C are in equilibrium. The systems A, B, and C are at the same temperature.

First Law of Thermodynamics

Many power plants and engines operate by turning heat energy into work. The reason is that a heated gas can do work on mechanical turbines or pistons, causing them to move. The first law of thermodynamics applies the conservation of energy principle to systems where heat transfer and doing work are the methods of transferring energy into and out of the system. The first law of thermodynamics states that the change in internal energy of a system ΔU equals the net heat transfer into the system Q, plus the net work done on the system W. In equation form, the first law of thermodynamics is,

$$\Delta U = Q + W$$

Here ΔU is the change in internal energy U of the system. Q is the net heat transferred into the system—that is, Q is the sum of all heat transfer into and out of the system. W is the net work done on the system.

So positive heat Q adds energy to the system and positive work W adds energy to the system. This is why the first law takes the form it does, $\Delta U = Q + W$. It simply says that you can add to the internal energy by heating a system, or doing work on the system.

Nothing quite exemplifies the first law of thermodynamics as well as a gas (like air or helium) trapped in a container with a tightly fitting movable piston. We'll assume the piston can move up and down, compressing the gas or allowing the gas to expand (but no gas is allowed to escape the container).

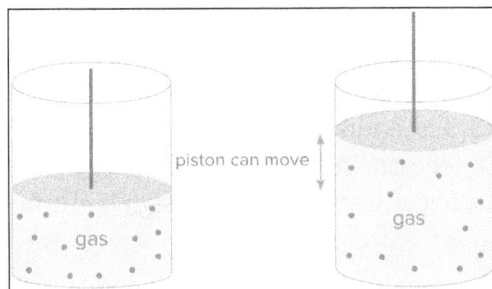

The gas molecules trapped in the container are the "system". Those gas molecules have kinetic energy.

The internal energy U of our system can be thought of as the sum of all the kinetic energies of the individual gas molecules. So, if the temperature T of the gas increases, the gas molecules speed up and the internal energy U of the gas increases (which means ΔU is positive). Similarly, if the temperature T of the gas decreases, the gas molecules slow down, and the internal energy U of the gas decreases (which means ΔU is negative).

It's really important to remember that internal energy U and temperature T will both increase when the speeds of the gas molecules increase, since they are really just two ways of measuring the same thing; how much energy is in a system. Since temperature and internal energy are proportional $T \propto U$, if the internal energy doubles the temperature doubles. Similarly, if the temperature does not change, the internal energy does not change.

One way we can increase the internal energy U (and therefore the temperature) of the gas is by transferring heat Q into the gas. We can do this by placing the container over a Bunsen burner or submerging it in boiling water. The high temperature environment would then conduct heat thermally through the walls of the container and into the gas, causing the gas molecules to move faster. If heat enters the gas, Q will be a positive number. Conversely, we can decrease the internal energy of the gas by transferring heat out of the gas. We could do this by placing the container in an ice bath. If heat exits the gas, Q will be a negative number. This sign convention for heat Q is represented in the image below.

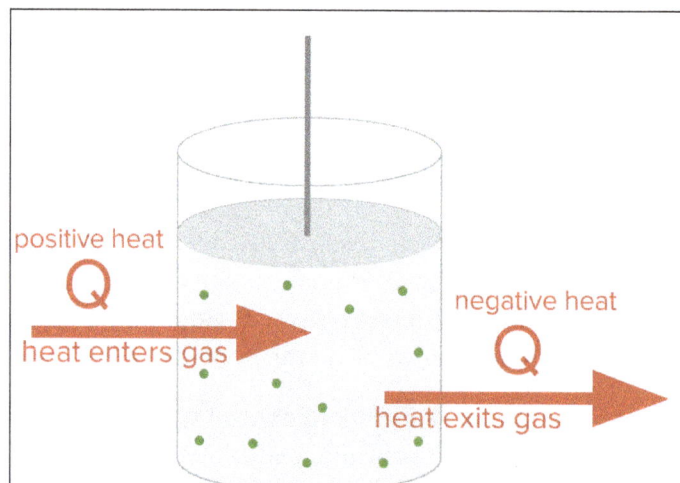

Since the piston can move, the piston can do work on the gas by moving downward and compressing the gas. The collision of the downward moving piston with the gas molecules causes the gas molecules to move faster, increasing the total internal energy. If the gas is compressed, the work done on the gas $W_{on\ gas}$ is a positive number. Conversely, if the gas expands and pushes the piston upward, work is done by the gas. The collision of the gas molecules with the receding piston causes the gas molecules to slow down, decreasing the internal energy of the gas. If the gas expands, the work done on the gas $W_{on\ gas}$ is a negative number. This sign convention for work W is represented in the image below.

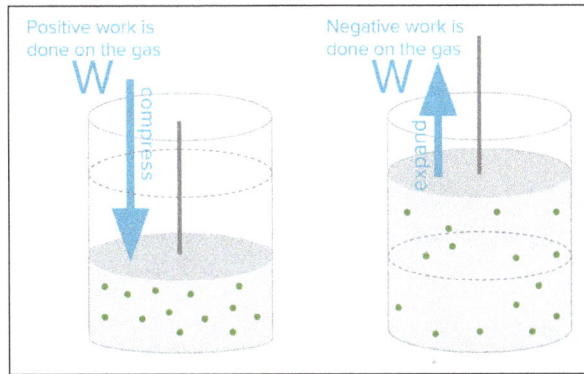

Below is a table that summarizes the signs conventions for all three quantities (ΔU, Q, W) discussed above.

ΔU (change in internal energy)	Q(heat)	W (work done on gas)
is + if temperature T increases	is + if heat enters gas	is + if gas is compressed
is – if temperature T decreases	is – if heat exits gas	is – if gas expands
is 0 if temperature T is constant	is 0 if no heat exchanged	is 0 if volume is constant

Heat Q and Temperature T

The heat Q represents the heat energy that enters a gas (e.g. thermal conduction through the walls of the container). The temperature T on the other hand, is a number that's proportional to the total internal energy of the gas. So, Q is the energy a gas gains through thermal conduction, but T is proportional to the total amount of energy a gas has at a given moment. The heat that enters a gas might be zero (Q = 0) if the container is thermally insulated, however, that does not mean that the temperature of the gas is zero (since the gas likely had some internal energy to start with).

To drive this point home, consider the fact that the temperature T of a gas can increase even if heat Q leaves the gas. This sounds counterintuitive, but since both work and heat can change the internal energy of a gas, they can both affect the temperature of a gas. For instance, if you place a piston in a sink of ice water, heat will conduct energy out the gas. However, if we compress the piston so that the work done on the gas is greater than the heat energy that leaves the gas, the total internal energy of the gas will increase.

Example: Nitrogen piston

A container has a sample of nitrogen gas and a tightly fitting movable piston that does not allow any of the gas to escape. During a thermodynamics process, 200 joules of heat enter the gas, and the gas does 300 joules of work in the process.

What was the change in internal energy of the gas during the process described above?

Solution:

We'll start with the first law of thermodynamics.

$\Delta U = Q + W$ (start with the first law of thermodynamics)

$$\Delta U = (+200\,J) + W (\text{plug in } Q = +200\,J)$$

$$\Delta U = (+200\,J) + (-300\,J)(\text{plug in } W = -300\,J)$$

$$\Delta U = -100\,J$$

Note: Since the internal energy of the gas decreases, the temperature must decrease as well.

Example: Heating helium

Four identical containers have equal amounts of helium gas that all start at the same initial temperature. Containers of gas also have a tightly fitting movable piston that does not allow any of the gas to escape. Each sample of gas is taken through a different process as described below:

Sample 1: 500 J of heat exits the gas and the gas does 300 J of work

Sample 2: 500 J of heat enters the gas and the gas does 300 J of work

Sample 3: 500 J of heat exits the gas and 300 J of work is done on the gas

Sample 4: 500 J of heat enters the gas and 300 J of work is done on the gas

Which of the following correctly ranks the final temperatures of the samples of gas after they're taken through the processes described above.

A. $T_4 > T_3 > T_2 > T_1$
B. $T_1 > T_3 > T_2 > T_4$
C. $T_4 > T_2 > T_3 > T_1$
D. $T_1 > T_4 > T_3 > T_2$

Solution:

Whichever gas has the largest increase in internal energy ΔU will also have the greatest increase in temperature ΔT (since temperature and internal energy are proportional). To determine how the internal energy changes, we'll use the first law of thermodynamics for each process.

Process 1:

$$\Delta U = Q + W$$

$$\Delta U = (-500\,J) + (-300\,J)$$

$$\Delta U = -800\,J$$

Process 2:

$$\Delta U = Q + W$$

$$\Delta U = (+500\,J) + (-300\,J)$$

$$\Delta U = +200\,J$$

Process 3:

$$\Delta U = Q + W$$
$$\Delta U = (-500\,J) + (300\,J)$$
$$\Delta U = -200\,J$$

Process 4:

$$\Delta U = Q + W$$
$$\Delta U = (+500\,J) + (+300\,J)$$
$$\Delta U = +800\,J$$

The final temperatures of the gas will have the same ranking as the changes in internal energy (i.e. sample 4 has the largest increase in internal energy, so sample 4 will end with the largest temperature).

$$\Delta U_4 > \Delta U_2 > \Delta U_3 > \Delta U_1 \text{ and } T_4 > T_2 > T_3 > T_1$$

So the correct answer is C.

Second Law of Thermodynamics

The Second Law of Thermodynamics states that the state of entropy of the entire universe, as an isolated system, will always increase over time. The second law also states that the changes in the entropy in the universe can never be negative.

To understand why entropy increases and decreases, it is important to recognize that two changes in entropy have to considered at all times. The entropy change of the surroundings and the entropy change of the system itself. Given the entropy change of the universe is equivalent to the sums of the changes in entropy of the system and surroundings:

$$\Delta S_{univ} = \Delta S_{sys} + \Delta S_{surr} = \frac{q_{sys}}{T} + \frac{q_{surr}}{T}$$

In an isothermal reversible expansion, the heat q absorbed by the system from the surroundings is,

$$q_{rev} = nRT \ln \frac{V_2}{V_1}$$

Since the heat absorbed by the system is the amount lost by the surroundings $q_{sys} = -q_{surr}$. Therefore, for a truly reversible process, the entropy change is,

$$\Delta S_{univ} = \frac{nRT \ln \dfrac{V_2}{V_1}}{T} + \frac{-nRT \ln \dfrac{V_2}{V_1}}{T} = 0$$

If the process is irreversible however, the entropy change is

$$\Delta S_{univ} = \frac{nRT \ln \frac{V_2}{V_1}}{T} > 0$$

If we put the two equations for ΔS_{univ} together for both types of processes, we are left with the second law of thermodynamics,

$$\Delta S_{univ} = \Delta S_{sys} + \Delta S_{surr} \geq 0$$

where ΔS_{univ} equals zero for a truly reversible process and is greater than zero for an irreversible process. In reality, however, truly reversible processes never happen (or will take an infinitely long time to happen), so it is safe to say all thermodynamic processes we encounter everyday are irreversible in the direction they occur.

The second law of thermodynamics can also be stated that "all spontaneous processes produce an increase in the entropy of the universe".

Gibbs Free Energy

Given another equation:

$$\Delta S_{total} = \Delta S_{univ} = \Delta S_{surr} + \Delta S_{sys}$$

The formula for the entropy change in the surroundings is $\Delta S_{surr} = \Delta H_{sys} / T$. If this equation is replaced in the previous formula, and the equation is then multiplied by T and by -1 it results in the following formula.

$$-T\Delta S_{univ} = \Delta H_{sys} - T\Delta S_{sys}$$

If the left side of the equation is replaced by G , which is know as Gibbs energy or free energy, the equation becomes

$$\Delta G = \Delta H - T\Delta S$$

Now it is much simpler to conclude whether a system is spontaneous, non-spontaneous, or at equilibrium.

- ΔH refers to the heat change for a reaction. A positive ΔH means that heat is taken from the environment (endothermic). A negative ΔH means that heat is emitted or given the environment (exothermic).

- ΔG is a measure for the change of a system's free energy in which a reaction takes place at constant pressure (P) and temperature (T).

According to the equation, when the entropy decreases and enthalpy increases the free energy change, ΔG , is positive and not spontaneous, and it does not matter what the temperature of the system is. Temperature comes into play when the entropy and enthalpy both increase or both decrease. The reaction is not spontaneous when both entropy and enthalpy are positive and at low temperatures, and the reaction is spontaneous when both entropy and enthalpy are positive and at high temperatures. The reactions are spontaneous when the entropy and enthalpy are negative at low temperatures, and the reaction is not spontaneous when the entropy and enthalpy are negative at high temperatures. Because all spontaneous reactions increase entropy, one can determine if the entropy changes according to the spontaneous nature of the reaction.

Third Law of Thermodynamics

The third law of thermodynamics states that the entropy of a perfect crystal at a temperature of zero Kelvin (absolute zero) is equal to zero.

Entropy, denoted by 'S', is a measure of the disorder/randomness in a closed system. It is directly related to the number of microstates (a fixed microscopic state that can be occupied by a system) accessible by the system, i.e. the greater the number of microstates the closed system can occupy, the greater its entropy. The microstate in which the energy of the system is at its minimum is called the ground state of the system.

At a temperature of zero Kelvin, the following phenomena can be observed in a closed system:

- The system does not contain any heat.

- All the atoms and molecules in the system are at their lowest energy points.

Therefore, a system at absolute zero has only one accessible microstate – it's ground state. As per the third law of thermodynamics, the entropy of such a system is exactly zero.

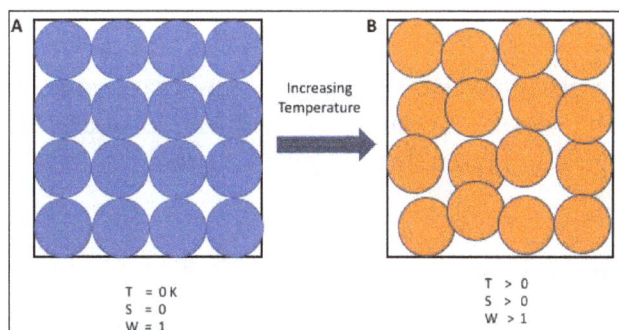

This law was developed by the German chemist Walther Nernst between the years 1906 and 1912.

Alternate Statements of the 3rd Law of Thermodynamics

The Nernst statement of the third law of thermodynamics implies that it is not possible for a process to bring the entropy of a given system to zero in a finite number of operations.

The American physical chemists Merle Randall and Gilbert Lewis stated this law in a different manner: when the entropy of each and every element (in their perfectly crystalline states) is taken as 0 at absolute zero temperature, the entropy of every substance must have a positive, finite value. However, the entropy at absolute zero can be equal to zero, as is the case when a perfect crystal is considered.

The Nernst-Simon statement of the 3rd law of thermodynamics can be written as: for a condensed system undergoing an isothermal process *that is reversible in nature, the associated entropy change approaches zero as the associated temperature approaches zero.*

Another implication of the third law of thermodynamics is: the exchange of energy between two thermodynamic systems (whose composite constitutes an isolated system) is bounded.

Reasons for not Achieving a Temperature at Zero Kelvins

For an isentropic process that reduces the temperature of some substance by modifying some parameter X to bring about a change from 'X$_2$' to 'X$_1$', an infinite number of steps must be performed in order to cool the substance to zero Kelvin.

This is because the third law of thermodynamics states that the entropy change at absolute zero temperatures is zero. The entropy v/s temperature graph for any isentropic process attempting to cool a substance to absolute zero is illustrated below.

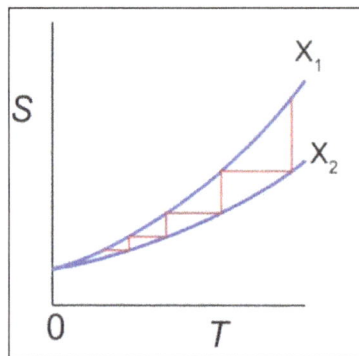

From the graph, it can be observed that – the lower the temperature associated with the substance, the greater the number of steps required to cool the substance further. As the temperature approaches zero kelvin, the number of steps required to cool the substance further approaches infinity.

Mathematical Explanation of the Third Law

As per statistical mechanics, the entropy of a system can be expressed via the following equation:

$S - So = kB \ln\Omega$

Where,

- S is the entropy of the system.

- S$_o$ is the initial entropy.

- k_B denotes the Boltzmann constant.

- Ω refers to the total number of microstates that are consistent with the system's macroscopic configuration.

Now, for a perfect crystal that has exactly one unique ground state, $\Omega = 1$. Therefore, the equation can be rewritten as follows:

$$S - S_0 = k_B \ln(1) = 0 \text{ [because } \ln(1) = 0\text{]}$$

When the initial entropy of the system is selected as zero, the following value of 'S' can be obtained:

$$S - 0 = 0 \Rightarrow S = 0$$

Thus, the entropy of a perfect crystal at absolute zero is zero.

Applications of the Third Law of Thermodynamics

An important application of the third law of thermodynamics is that it helps in the calculation of the absolute entropy of a substance at any temperature 'T'. These determinations are based on the heat capacitymeasurements of the substance. For any solid, let S_0 be the entropy at 0 K and S be the entropy at T K, then

$$\Delta S = S - S_0 = \int_0^T \frac{C_p}{T} dT$$

According to the third law of thermodynamics, $S_0 = 0$ at 0 K,

$$S = \int_0^T \frac{C_p}{T} dT$$

The value of this integral can be obtained by plotting the graph of C_p / T versus T and then finding the area of this curve from 0 to T. The simplified expression for the absolute entropy of a solid at temperature T is as follows:

$$S = \int_0^T \frac{C_p}{T} dT = \int_0^T C_p \, d \ln T$$

$$= C_p \ln T = 2.303 \, C_p \log T$$

Here C_p is the heat capacity of the substance at constant pressure and this value is assumed to be constant in the range of 0 to T K.

Entropy

Entropy is the measure of a system's thermal energy per unit temperature that is unavailable for doing useful work. Because work is obtained from ordered molecular motion, the amount of entropy is also a measure of the molecular disorder, or randomness, of a system. The concept of entropy provides deep insight into the direction of spontaneous change for many everyday phenomena. Its introduction by the German physicist Rudolf Clausius in 1850 is a highlight of 19th-century physics.

The idea of entropy provides a mathematical way to encode the intuitive notion of which processes are impossible, even though they would not violate the fundamental law of conservation of energy. For example, a block of ice placed on a hot stove surely melts, while the stove grows cooler. Such a process is called irreversible because no slight change will cause the melted water to turn back into ice while the stove grows hotter. In contrast, a block of ice placed in an ice-water bath will either thaw a little more or freeze a little more, depending on whether a small amount of heat is added to or subtracted from the system. Such a process is reversible because only an infinitesimal amount of heat is needed to change its direction from progressive freezing to progressive thawing. Similarly, compressed gas confined in a cylinder could either expand freely into the atmosphere if a valve were opened (an irreversible process), or it could do useful work by pushing a moveable piston against the force needed to confine the gas. The latter process is reversible because only a slight increase in the restraining force could reverse the direction of the process from expansion to compression. For reversible processes the system is in equilibrium with its environment, while for irreversible processes it is not.

To provide a quantitative measure for the direction of spontaneous change, Clausius introduced the concept of entropy as a precise way of expressing the second law of thermodynamics. The Clausius form of the second law states that spontaneous change for an irreversible process in an isolated system (that is, one that does not exchange heat or work with its surroundings) always proceeds in the direction of increasing entropy. For example, the block of ice and the stove constitute two parts of an isolated system for which total entropy increases as the ice melts.

By the Clausius definition, if an amount of heat Q flows into a large heat reservoir at temperature T above absolute zero, then the entropy increase is $\Delta S = Q/T$. This equation effectively gives an alternate definition of temperature that agrees with the usual definition. Assume that there are two heat reservoirs R_1 and R_2 at temperatures T_1 and T_2 (such as the stove and the block of ice). If an amount of heat Q flows from R_1 to R_2, then the net entropy change for the two reservoirs is

$$\Delta S = Q\left(\frac{1}{T_2} - \frac{1}{T_1}\right).$$

which is positive provided that $T_1 > T_2$. Thus, the observation that heat never flows spontaneously from cold to hot is equivalent to requiring the net entropy change to be positive for a spontaneous flow of heat. If $T_1 = T_2$, then the reservoirs are in equilibrium, no heat flows, and $\Delta S = 0$.

The condition $\Delta S \geq 0$ determines the maximum possible efficiency of heat engines—that is, systems such as gasoline or steam engines that can do work in a cyclic fashion. Suppose a heat engine absorbs heat Q_1 from R_1 and exhausts heat Q_2 to R_2 for each complete cycle. By conservation of energy, the work done per cycle is $W = Q_1 - Q_2$, and the net entropy change is

$$\Delta S = \frac{Q_2}{T_2} - \frac{Q_1}{T_1}$$

To make W as large as possible, Q_2 should be as small as possible relative to Q_1. However, Q_2 cannot be zero, because this would make ΔS negative and so violate the second law. The smallest possible value of Q_2 corresponds to the condition $\Delta S = 0$, yielding

$$\left(\frac{Q_2}{Q_1}\right)_{min} = \frac{T_2}{T_1}$$

as the fundamental equation limiting the efficiency of all heat engines. A process for which $\Delta S = 0$ is reversible because an infinitesimal change would be sufficient to make the heat engine run backward as a refrigerator.

The same reasoning can also determine the entropy change for the working substance in the heat engine, such as a gas in a cylinder with a movable piston. If the gas absorbs an incremental amount of heat dQ from a heat reservoir at temperature T and expands reversibly against the maximum possible restraining pressure P, then it does the maximum work dW = PdV, where dV is the change in volume. The internal energy of the gas might also change by an amount dU as it expands. Then by conservation of energy, dQ = dU + P dV. Because the net entropy change for the system plus reservoir is zero when maximum work is done and the entropy of the reservoir decreases by an amount $dS_{reservoir} = -dQ/T$, this must be counterbalanced by an entropy increase of

$$DS_{system} = \frac{dU + p\,dV}{T} = \frac{dQ}{T}$$

for the working gas so that $dS_{system} + dS_{reservoir} = 0$. For any real process, less than the maximum work would be done (because of friction, for example), and so the actual amount of heat dQ' absorbed from the heat reservoir would be less than the maximum amount dQ. For example, the gas could be allowed to expand freely into a vacuum and do no work at all. Therefore, it can be stated that

$$dS_{system} = \frac{dU + p\,dV}{T} \geq \frac{dQ'}{T}.$$

with $dQ' = dQ$ in the case of maximum work corresponding to a reversible process.

This equation defines S_{system} as a thermodynamic state variable, meaning that its value is completely determined by the current state of the system and not by how the system reached that state. Entropy is an extensive property in that its magnitude depends on the amount of material in the system.

In one statistical interpretation of entropy, it is found that for a very large system in thermodynamic equilibrium, entropy S is proportional to the natural logarithm of a quantity Ω representing the maximum number of microscopic ways in which the macroscopic state corresponding to S can be realized; that is, S = k ln Ω, in which k is the Boltzmann constant that is related to molecular energy.

All spontaneous processes are irreversible; hence, it has been said that the entropy of the universe is increasing: that is, more and more energy becomes unavailable for conversion into work. Because of this, the universe is said to be "running down."

Statistical Mechanics

Statistical mechanics describes the thermodynamic behaviour of macroscopic systems from the laws which govern the behaviour of the constituent elements at the microscopic level. The microscopic elements can be atoms, molecules, dipole moments or magnetic moments, etc. A macroscopic system, generally, is composed of a large number of these elements (of the order of Avogadro number $N_A \approx 6.022 \times 10^{23}$ per mole). Each element may have a large number of internal degrees of freedom associated with different types of motion such as translation, rotation, vibration etc. The constituent elements may interact with the external field applied to the system. There can also be very complex interaction among the constituent elements. The macroscopic properties of a system is thus determined in the thermodynamic limit by the properties of the constituent molecules, their interaction with external field as well as interaction among themselves.

Specification of Macrostates and Microstates

Macrostate:The macroscopic state of a thermodynamic system at equilibrium is specified by the values of a set of measurable thermodynamic parameters. For example, the macrostate of a fluid system can be specified by pressure, temperature and volume, (P,V,T) . For an isolated system for which there is no exchange of energy or mass with the surroundings, the macrostate is specified by the internal energy E , number of particles N and volume V ; (E,N,V) . In case of a closed system which exchanges only energy with the surroundings (and no particle) and is in thermodynamic equilibrium with a heat bath at temperature T , the macrostate is given by (N,V,T) . For an open system, exchange of both energy and particle with the surroundings can take place. For such systems, at equilibrium with a heat bath at temperature T and a pressure bath at pressure P (or a particle bath of chemical potential μ), the macrostate is specified by (N,P,T) (or (μ,V,T))

The equilibrium of an isolated system corresponds to maximum entropy S(E,N,V) , for a closed system it corresponds to minimum Helmholtz free energy F(N,V,T) , and for an open system the equilibrium corresponds to minimum of Gibb's free energy G(N,P,T) (or minimum of the grand potential $\Phi(\mu,V.T)$).

Microstate: A microstate of a system is obtained by specifying the states of all of its constituent elements. However, it depends on the nature of the constituent elements (or particles) of the system. Specification of microstates are made differently for classical and quantum particles.

Microstates of classical particles: In order to specify the microstates of a system of classical particles, one needs to specify the position q and the conjugate momentum p of each and every constituent particle of the system. In a classical system, the time evolution of q and p is governed by the classical Hamiltonian H(p,q) and Hamilton's equation of motion

$$\dot{q}_i \frac{\partial H(p,q)}{\partial p_i} \quad and \quad \dot{p}_i = -\frac{\partial H(p,q)}{\partial q_i}; i = 1, 2, ..., 3N$$

For a system of N particles in 3-dimension. The state of a single particle at any time is then given by the pair of conjugate variables (qi , pi), a point in the phase space. Each single particle then constitutes a 6 -dimensional phase space (3 -coordinate and 3 -momentum). For N particles, the state of the system is then completely and uniquely defined by 3N canonical coordinates q_1, q_2, q_{3N} , and 3N canonical momenta $p_1, p_2, ... p_{3N}$. These 6N variables constitute a 6N -dimensional Γ -space or phase space of the system and each point of the phase space represents a microstate of the system. The locus of all the points in Γ -space satisfying the condition H(p,q) = E , total energy of the system, defines the energy surface.

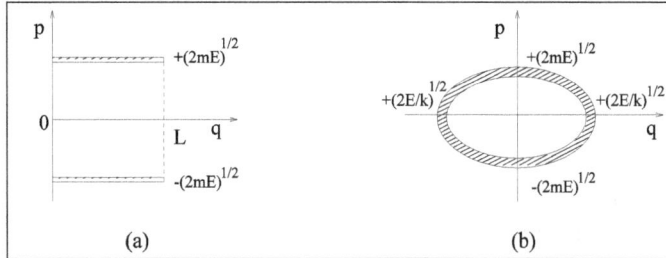

Figure: (a) Accessible region of phase space for a free particle of mass m and energy E in a one dimensional box of length L . (b) Region of phase space for a one dimensional harmonic oscillator with energy E , mass m and spring constant k .

Example: Consider a free particle of mass m inside a one dimensional box of length L , such that $0 \le q \le L$, with energy between E and E +δE . The macroscopic state of the system is defined by (E,N, L) with N =1 . The microstates are specified in certain region of phase space. Since the energy of the particle is E = $p^2/2m$, the momentum will be $p = \pm\sqrt{2mE}$ and the position q is within 0 and L. However, there is a small width in energy δE, so the particles are confined in small strips of width $\delta p = \sqrt{m/2E}\delta E$ as shown in Fig. (a). Note that if δE = 0 , the accessible region of phase space representing the system would be one dimensional in a two dimensional phase space. In order to avoid this artifact a small width in E is considered which does not affect the final results in the thermodynamic limit. In Fig. (b), the phase space region of a one dimensional harmonic oscillator with mass m , spring constant k and energy between E and E +δE is shown. The Hamiltonian of the particle is: $H = p^2/2m + kq^2/2$ and for a given energy E , the accessible region is an ellipse: $p^2/(2mE) + q^2/(2E/k) = 1$. With the energy between E and E +δE , the accessible region is an elliptical shell of area $2\pi\sqrt{m/k}\delta E$.

Microstates of quantum particles: For a quantum particle, the state is characterized by the wave function Ψ (q_1 , q_2, q_3 ,) . Generally, the wave function is written in terms of a complete orthonormal basis of eigenfunctions of the Hamiltonian operator of the system. Thus, the wave function may be written as

$$\Psi = \sum_n c_n \phi_n, \quad \hat{H}\phi_n = E_n\phi_n$$

where E_n is the eigenvalue corresponding to the state ϕ_n. The eigenstates ϕ_n, characterized by a set of quantum numbers n provides a way to count the microscopic states of the system.

Example: Consider a localized magnetic ion of spin 1/2 and magnetic moment μ in thermal equilibrium at temperature T . The particle has two eigenstates, (1,0) and (0,1) associated with spin up ($\uparrow$) and down spin ($\downarrow$) respectively. In the presence of an external magnetic field $\vec{H}$, the energy is given by

$$E = -\vec{\mu}.\vec{H} = \begin{cases} +\mu H, \textit{for spin} \downarrow \\ -\mu H, \textit{for spin} \uparrow \end{cases}$$

Thus, the system with macrostate (N,H,T) with N =1 has two microstates with energy $-\mu H$ and $+\mu H$ corresponding to up spin (parallel to H ρ) and down spin (antiparallel to H ρ). If there are two such magnetic ions in the system, it will have four microstates: $\uparrow\uparrow$ with energy $-2\mu H$, $\uparrow\downarrow$ &$\downarrow\uparrow$ with zero energy and $\downarrow\downarrow$ with energy $+2\mu H$. For a system of N spins of spin -1/2 , there are total N 2 microstates and specification of the spin-states of all the N spins will give one possible microstate of the system.

Statistical Ensembles

An ensemble is a collection of a large number of replicas (or mental copies) of the microstates of the system under the same macroscopic condition or having the same macrostate. However, the microstates of the members of an ensemble can be arbitrarily different. Thus, for a given macroscopic condition, a system of an ensemble is represented by a point in the phase space. The ensemble of a macroscopic system of given macrostate then corresponds to a large number of points in the phase space. During time evolution of a macroscopic system in a fixed macrostate, the microstate is supposed to pass through all these phase points.

Depending on the interaction of a system with the surroundings (or universe), a thermodynamic system is classified as isolated, closed or open system. Similarly, statistical ensembles are also classified into three different types. The classification of ensembles again depends on the type of interaction of the system with the surroundings which can either be by exchange of energy only or exchange of both energy and matter (particles or mass). In an isolated system, neither energy nor matter is exchanged and the corresponding ensemble is known as microcanonical ensemble. A closed system exchanging only energy (not matter) with its surroundings is described by canonical ensemble. Both energy and matter are exchanged between the system and the surroundings in an open system and the corresponding ensemble is called a grand canonical ensemble.

Statistical Equilibrium

Consider an isolated system with the macrostate (E,N,V). A point in the phase space corresponds to a microstate of such a system and its internal dynamics is described by the corresponding phase trajectory. The density of phase points $\rho(p,q)$ is the number of microstates per unit volume of the phase space and it is the probability to find a state around a phase point (p,q).

By Liouville's theorem, in the absence of any source and sink in the phase space, the total time derivative in the time evolution of the phase point density $\rho(p,q)$ is given by

$$\frac{d\rho}{dt} = \frac{\partial\rho}{\partial t} + \{\rho, H\} = 0$$

Where

$$\{\rho, H\} = \sum_{i=1}^{3N}\left(\frac{\partial\rho}{\partial q_i}\frac{\partial H}{\partial p_i} - \frac{\partial\rho}{\partial p_i}\frac{\partial H}{\partial q_i}\right)$$

is the Poisson bracket of the density function ρ and the Hamiltonian H of the system.

The ensemble is considered to be in statistical equilibrium if ρ(p,q) has no explicit dependence on time at all points in the phase space, i.e.,

$$\frac{\partial \rho}{\partial t} = 0.$$

Under the condition of equilibrium, therefore, {ρ, H}= 0 . It will be satisfied if ρ is an explicit function of the Hamiltonian H(q, p) or ρ is a constant, independent of p and q . That is

$$\rho(p,q) = \text{constant}.$$

The condition of statistical equilibrium then requires no explicit time dependence of the phase point density ρ(p,q) as well as uniform distribution of ρ(p,q) over the relevant region of phase space. The value of ρ(p,q) will, of course, be zero outside the relevant region of phase space. Physically the choice corresponds to an ensemble of systems which at all times are uniformly distributed over all possible microstates and the resulting ensemble is referred to as the microcanonical ensemble. However, in canonical ensemble it can be shown that ρ(q, p) ∝ exp[–H(q, p)/k_BT].

Postulates of Statistical Mechanics

The principles of statistical mechanics and their applications are based on the following two postulates.

Equal a priori probability: For a given macrostate (E,N,V), specified by the number of particles N in the system of volume V and at energy E, there is usually a large number of possible microstates of the system. In case of classical non-interacting system, the total energy E can be distributed among the N particles in a large number of different ways and each of these different ways corresponds to a microstate. In the fixed energy ensemble, the density ρ(q, p) of the representative points in the phase space corresponding to these microstates is constant or the phase points are uniformly distributed. Thus, any member of the ensemble is equally likely to be in any of the various possible microstates. In case of a quantum system, the various different microstates are identified as the independent solutions $\Psi(r_1, r_2, ..., r_N)$ of the Schrödinger equation of the system, corresponding to an eigenvalue E . At any time t , the system is equally likely to be in any one of these microstates. This is generally referred as the postulate of equal a priori probability for all microstates of a given macrostate of the system.

Principle of ergodicity: The microstates of a macroscopic system are specified by a set of points in the 6N -dimensional phase space. At any time t , the system is equally likely to be in any one of the large number of microstates corresponding to a given macrostate, say (E,N,V) as for an isolated system. With time, the system passes from one microstate to another. After a sufficiently long time, the system passes through all its possible microstates. In the language of statistical mechanics, the system is considered to be in equilibrium if it samples all the microstates with equal a priori probability. The equilibrium value of the observable X can be obtained by the statistical or ensemble average

$$\langle X \rangle = \frac{\iint X(p,q)\rho(p,q)d^{3N}qd^{3N}p}{\int \rho(p,q)d^{3N}qd^{3N}p}.$$

On the other hand, the mean value of an observable (or a property) is given by its time-averaged value:

$$\bar{X} = \lim_{T \to \infty} \frac{1}{T} \int_0^T X(t)dt.$$

The ergodicity principle suggests that statistical average $\langle X \rangle$ and the mean value $\bar{X}$ are equivalent: $\bar{X} \equiv \langle X \rangle$.

Reference

- Heat, thermodynamics-chemistry, chemistry, science: khanacademy.org, Retrieved 24 April, 2019

- Thermodynamics: newworldencyclopedia.org, Retrieved 27 August, 2019

- The-zeroth-law-of-thermodynamics, boundless-physics: lumenlearning.com, Retrieved 21 June, 2019

- What-is-the-first-law-of-thermodynamics, laws-of-thermodynamics, thermodynamics, physics, science: khanacademy.org, Retrieved 11 May, 2019

- Second_Law_of_Thermodynamics, The_Four_Laws_of_Thermodynamics, Thermodynamics, Physical_ and_Theoretical_Chemistry: libretexts.org, Retrieved 20 July, 2019

- Third-law-of-thermodynamics, chemistry: byjus.com, Retrieved 15 March, 2019

- Entropy-physics, science: britannica.com, Retrieved 4 January, 2019

Electromagnetic Theory

Electromagnetic deals with the study of electromagnetic force, which is a type of physical interaction that occurs between electrically charged particles. It is concerned with the study of effects related to charge such as electric current, electric field, electric potentials, etc. This chapter discusses in detail the theories related to electromagnetic theory.

Electric Charge

Electric charge is the property of sub-atomic particles particularly includes electrons and protons. Electrons have negative charge, protons have positive charge and neutrons do not have any charge.

There are two types of electric charges. They are positive charges and negative charges.

Charges are present in almost every type of body. All those bodies having no charges are the neutrally charged ones. We denote a charge y the symbol 'q' and its standard unit is Coulomb. Mathematically, we can say that a charge is the number of electrons multiplied by the charge on 1 electron. Symbolically, it is

$$Q = ne$$

where q is a charge, n is a number of electrons and e is a charge on 1 electron ($1.6 \times 10\text{-}19$C). The two very basic natures of electric charges are

- Like charges repel each other.
- Unlike charges attract each other.

This means that while protons repel protons, they attract electrons. The nature of charges is responsible for the forces acting on them and coordinating the direction of the flow of them. The charge on electron and proton is the same in magnitude which is $1.6 \times 10\text{-}19$ C. The difference is only the sign that we use to denote them, + and -.

Basic Properties of Electric Charge

There are certain other basic properties that an electric charge follows from the electric charge definition. They are

- Charges are additive in nature
- A charge is a conserved quantity
- Quantization of charge

Charges as Additive in Nature

This means that they behave like scalars and we can add them directly. As an example, let us consider a system which consists of two charges namely q_1 and q_2. The total charge of the system will be the algebraic sum of q_1 and q_2 i.e., $q_1 + q_2$. The same thing holds for a number of charges in a system. Let's say a system contains $q_1, q_2, q_3, q_4 \ldots \ldots \ldots q_n$ then the net charge of the entire system will be

$$= q_1 + q_2 + q_3 + q_4 + \ldots \ldots + q_n$$

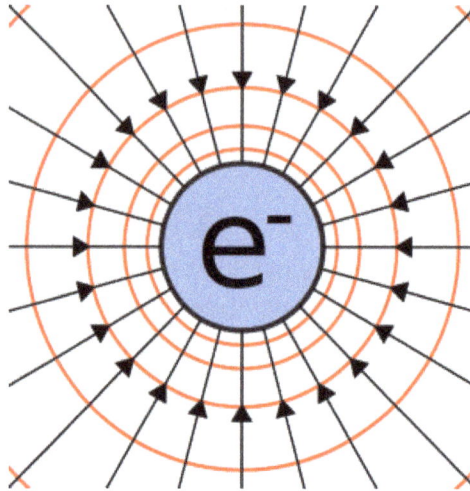

Charge as Conserved Quantity

This implies that charge can neither be created nor be destroyed but can be transferred from one body to another by certain methods like conduction and induction. Does this remind you of the law of conservation of mass? As charging involves rubbing two bodies, it is actually a transfer of electrons from one body to another.

For example, if 5 C is the total charge of the system, then we can redistribute it as 1C, 2C, and 2C or in any other possible permutation. For example, sometimes a neutrino decays to give one electron and one proton by default in nature. The net charge of the system will be zero as electrons and protons are of the same magnitude and opposite signs.

Quantization of Charge

This signifies the fact that charge is a quantized quantity and we can express it as integral multiples of the basic unit of charge (e – charge on one electron). Suppose charge on a body is q, then we can write it as,

$$q = ne$$

where n is an integer and not fraction or irrational number, like 'n' can be any positive or negative integer like 1, 2, 3, -5, etc. The basic unit of charge is the charge that an electron or proton carries. By convention, we take charge of the electron as negative and denote it as "$-e$" and charge on a proton is simply "e".

English experimentalist Faraday was the first to propose the quantization of charge principle. He did this when he put forward his experimental laws of electrolysis. Millikan in 1912, finally demonstrated and proved this principle.

1 A Coulomb of charge contains around 6×10^{18} electrons. Particles don't have a high magnitude of charge and we use micro coulombs or milli coulombs in order to express charge of a particle.

$$1\,\mu C = 10^{-6}\,C$$
$$1\,mC = 10^{-3}\,C$$

We can use the principle of quantization to calculate the total amount of charge present in a body and also to calculate a number of electrons or protons in a body. Suppose a system has n_1 number of electrons and n_2 number of protons, then the total amount of charge will be $n_2 e - n_1 e$.

Electric Current

Electric current is the movement of electric charge carriers, such as subatomic charged particles (e.g., electrons having negative charge, protons having positive charge), ions (atoms that have lost or gained one or more electrons), or holes (electron deficiencies that may be thought of as positive particles).

Electric current in a wire, where the charge carriers are electrons, is a measure of the quantity of charge passing any point of the wire per unit of time. In alternating current the motion of the electric charges is periodically reversed; in direct current it is not. In many contexts the direction of the current in electric circuits is taken as the direction of positive charge flow, the direction opposite to the actual electron drift. When so defined the current is called conventional current.

Current is usually denoted by the symbol I.

Unit of Current

Since the charge is measured in coulombs and time in seconds, so the unit of electric current is coulomb/Sec (C/s) or amperes (A). The amperes is the SI unit of the conductor. The I is the symbolic representation of the current.

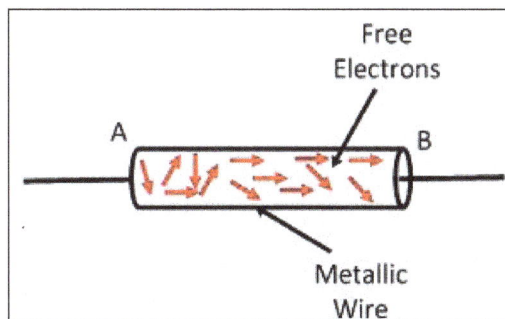

$Q = 1$ coulomb; $t = 1$ second then $I = 1A$

Thus, a wire is said to carry a current of one ampere when charge flows through it at the rate of one coulomb per second.

When an electrical potential difference is applied across the metallic wire, the loosely attached free electrons start moving towards the positive terminal of the cell shown in the figure below. This continuous flow of electrons constitutes the electrical current. The flow of currents in the wire is from the negative terminal of the cell to the positive terminal through the external circuit.

Conventional Direction of Flow of Current

According to the electron theory, when the potential difference is applied across the conductor some matter flows through the circuit which constitutes the electric current. It was considered that this matter flows from higher potential to lower potential, i.e. positive terminal to the negative terminal of the cell through the external circuit.

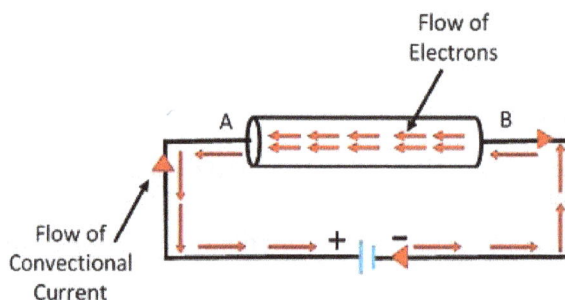

This convention of flow of current is so firmly established that it is still in use. Thus, the conventional direction of flow of current is from the positive terminal of the cell to the negative terminal of the cell through the external circuit. The magnitude of flow of current at any section of the conductor is the rate of flow of electrons i.e. charge flowing per second.

Mathematically, it is represented by:

$$I = Q/t$$

On the basis of the flow of electric charge the current is mainly classified into two types, i.e. alternating current and direct current. In direct current, the charges flow through unidirectional whereas in alternating current the charges flows in both the direction.

Effects of Current

When an electric current flows through a conductor there are a number of signs which tell that a current is flowing.

- Heat is dissipated: Possibly the most obvious is that heat is generated. If the current is small then the amount of heat generated is likely to be very small and may not be noticed. However if the current is larger then it is possible that a noticeable amount of heat is generated. An electric fire is a prime example showing how a current causes heat to be generated. The actual amount of heat is governed not only be the current, but also be the voltage and the resistance of the conductor.

- Magnetic effect: Another effect which can be noticed is that a magnetic field is built up around the conductor. If a current is flowing in conductor then it is possible to detect this. By placing a compass close to a wire carrying a reasonably large direct current, the compass needle can be seen to be deflect. Note this will not work with mains because the field is alternating too fast for the needle to respond and the two wires (live and neutral) close together in the same cable will cancel out the field.

The magnetic field generated by a current is put to good use in a number of areas. By winding a wire into a coil, the effect can be increased, and an electro-magnet can be made. Relays and a host of other items use the effect. Loudspeakers also use a varying current in a coil to cause vibrations to occur in a diaphragm which enable the electronic currents to be converted into sounds.

Electric Current and Circuit Diagram Elements

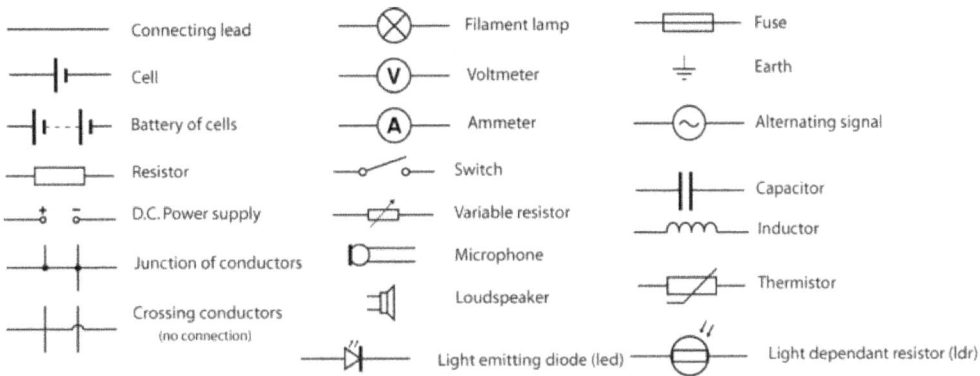

These symbols represent the common electrical components.

Example: A current of 0.75 A is drawn by the filament of an electric bulb for 10 minutes. Find the amount of electric charge that flows through the circuit.

Sol: Given, $I = 0.75\ A$, $t = 10\ minutes = 600\ s$

We know, $Q = I \times t = 0.75 \times 600$

Therefore, $Q = 45°C$

Electric Force

The repulsive or attractive interaction between any two charged bodies is called as an electric force. Similar to any force, its impact and effects on the given body are described by Newton's laws of motion. The electric force is among the list of other forces that exert over objects.

Newton's laws are applicable to analyze the motion under the influence of that kind of force or combination of forces. The analysis begins by the construction of a free body image wherein the direction and type of the individual forces are shown by the vector to calculate the resultant sum which is called as the net force that can be applied to determine the body's acceleration.

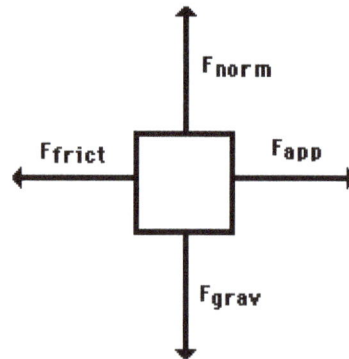

The electric force between two electrons is equal to the electric force between two protons when placed at equal distances. This describes that the electric force is not based on the mass of the object, but depends on the quantity known as the electric charge.

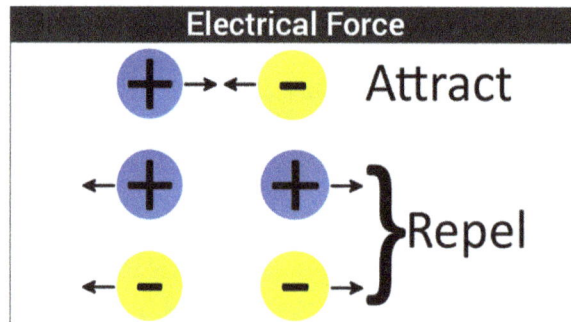

Coulomb's Law

Coulomb's Law gives an idea about the force between two point charges. By the word point charge, we mean that in physics, the size of linear charged bodies is very small as against the distance between them. Therefore, we consider them as point charges as it becomes easy for us to calculate the force of attraction/ repulsion between them.

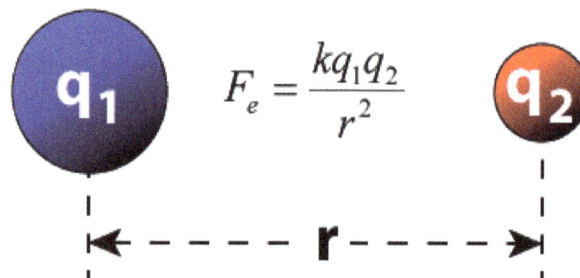

$$F_e = \frac{kq_1 q_2}{r^2}$$

Charles-Augustin de Coulomb, a French physicist in 1784, measured the force between two point charges and he came up with the theory that the force is inversely proportional to the square of the distance between the charges. He also found that this force is directly proportional to the product of charges (magnitudes only).

We can show it with the following explanation. Let's say that there are two charges q_1 and q_2. The distance between the charges is 'r', and the force of attraction/repulsion between them is 'F'. Then

$$F \propto q_1 q_2$$

Or, $F \propto 1/r^2$

$$F = k \, q_1 q_2 / r^2$$

where k is proportionality constant and equals to $1/4\pi\varepsilon_0$. Here, ε_0 is the epsilon naught and it signifies permittivity of a vacuum. The value of k comes $9 \times 10^9 \ Nm^2 / C^2$ when we take the *S.I.* unit of value of ε_0 *is* $8.854 \times 10^{-12} \ C^2 \ N^{-1} \ m^{-2}$.

According to this theory, like charges repel each other and unlike charges attract each other. This means charges of same sign will push each other with repulsive forces while charges with opposite signs will pull each other with attractive force.

Vector Form of Coulomb's Law

The physical quantities are of two types namely scalars (with the only magnitude) and vectors (those quantities with magnitude and direction). Force is a vector quantity as it has both magnitude and direction. The Coulomb's law can be re-written in the form of vectors. Remember we denote the vector "F" as F, vector r as r and so on.

Let there be two charges q_1 and q_2, with position vectors r_1 and r_2 respectively. Now, since both the charges are of the same sign, there will be a repulsive force between them. Let the force on the q_1 charge due to q_2 be F_{12} and force on q_2 charge due to q1 charge be F_{21}. The corresponding vector from q_1 to q_2 is r_{21} vector.

$$r_{21} = r_2 - r_1$$

To denote the direction of a vector from position vector r_1 to r_2, and from r_2 to r_1 as:

$$\widehat{r_{21}} = \frac{r_{21}}{r_{21}}.\hat{r}_{12} \frac{r_{12}}{r_{12}}.\widehat{r_{21}} = \widehat{r_{12}}$$

Now, the force on charge q_2 due to q_1, in vector form is:

$$F_{21} = \frac{1}{4\pi\varepsilon_0} \frac{q_1 q_2}{r_{21}^2} \hat{r}_{21}$$

The above equation is the vector form of Coulomb's Law.

Remarks on Vector Form of Coulomb's Law

While applying Coulomb's Law to find out the force between two point charges, we have to be careful of the following remarks. The vector form of the equation is independent of signs of both the charges, as both the forces are opposite in nature.

The repulsive force F_{12}, that is the force on charge q_1 due to q_2 and another repulsive force F_{21} that is the force on charge q_2 due to q_1 are opposite in signs, due to change in position vector.

$$F_{12} = -F_{21}$$

This is because the position vector in case of force F_{12} is r_{12} and position vector in case of force F_{21} is r_{21}, now

$$r_{21} = r_2 - r_1$$

$$r_{12} = r_1 - r_2$$

Since both r_{21} and r_{12} are opposite in signs, they make forces of opposite signs too. This proves that Coulomb's Law fits into Newton's Third Law i.e. every action has its equal and opposite reaction. Coulomb's Law provides the force between two charges when they're present in a vacuum. This is because charges are free in a vacuum and don't get interference from other matter or particles.

Limitations of Coulomb's Law

Coulomb's Law is derived under certain assumptions and can't be used freely like other general formulas. The law is limited to following points:

- We can use the formula if the charges are static (in rest position)

- The formula is easy to use while dealing with charges of regular and smooth shape, and it becomes too complex to deal with charges having irregular shapes

- The formula is only valid when the solvent molecules between the particle are sufficiently larger than both the charges.

Worked Example: Coulomb's Law

Two point-like charges carrying charges of $+3 \times 10^{-9}$ C *and* -5×10^{-9} C are 2 *m* apart. Determine the magnitude of the force between them and state whether it is attractive or repulsive.

Solution

Step 1: Determine what is required

We are required to determine the force between two point charges given the charges and the distance between them.

Step 2: Determine how to approach the problem

We can use Coulomb's law to calculate the magnitude of the force.

$$F = k \frac{Q_1 Q_2}{r^2}$$

Step 3: Determine what is given

We are given:

- $Q_1 = +3 \times 10^{-9} \ C$
- $Q_2 = -5 \times 10^{-9} \ C$
- $r = 2 \ m$

We know that $k = 9.0 \times 10^9 \ N \cdot m^2 \cdot C^{-2}$.

We can draw a diagram of the situation.

$$Q_1 = +3 \times 10^{-9} \ C \qquad\qquad Q_2 = -5 \times 10^{-9} \ C$$

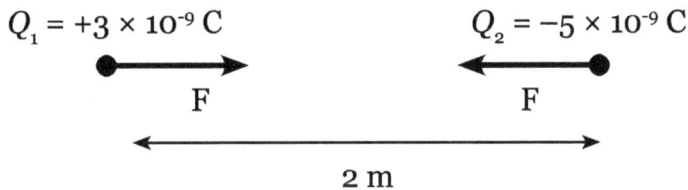

$$F \qquad\qquad\qquad\qquad F$$

$$2 \ m$$

Step 4: Check units

All quantities are in *SI* units.

Step 5: Determine the magnitude of the force

Using Coulomb's law we have

$$F = \frac{kQ_1Q_2}{r^2}$$
$$= \frac{(9.0 \times 10^9)(3 \times 10^{-9})(5 \times 10^{-9})}{(2)^2}$$
$$= 3.37 \times 10^{-8} \ N$$

Thus the magnitude of the force is 3.37×10⁻⁸ N. However since the point charges have opposite signs, the force will be attractive.

Step 6: Free body diagram

We can draw a free body diagram to show the forces. Each charge experiences a force with the same magnitude and the forces are attractive, so we have:

Principle of Superposition

Coulomb's law applies to any pair of point charges. When more than two charges are present, the net force on any one charge is simply the vector sum of the forces exerted on it by the other charges. For example, if three charges are present, the resultant force experienced by q_3 due to q_1, q_2 and will be

$$\vec{F}_3 = \vec{F}_{13} + \vec{F}_{23}$$

Electric Field

The electric field $\vec{E}$ is a vector quality that exists at every point in space. The electric field at a location indicates the force that would act on a unit positive test charge if placed at that location.

The electric field is related to the electronic force that acts on an arbitrary charge q by.

$$\vec{E} = \frac{\vec{F}}{q}$$

The dimension of electric field are newton's/coulomb, N/C.

We can express the electric force in terms of electric field,

$$\vec{F} = q\vec{E}$$

For a positive q, the electric field vector points in the same direction as the force vector.

The equation for electric field is similar to Coulomb's Law. We assign one of the $q's$ in the numerator of Coulomb's Law to play the role of test charge. The other charge(s) in the numerator, qi create the electric field we want to study.

Coulomb's Law: $\vec{F} = \frac{1}{4\pi \in_0} \frac{qqi}{r^2} \hat{r_i}$ newtons

Electric Field: $\vec{E} = \frac{\vec{F}}{q} = \frac{1}{4\pi \in_0} \frac{qi}{r^2} \hat{r_i}$ newtons/ coulomb

Where $\hat{r_i}$ are unit vectors indicating the line between each qi and q.

The electric field is normalized electric force. Electric field is the force experienced by a test charge that has a value of +1+1plus, 1.

One way to visualize the electric field (this is my mental model): imagined small positive test charge glued to the end of an imaginary stick. (Be sure your imaginary stick doesn't conduct, like wood or plastic). Explore the electric field by holding your test charge in various locations. The test charge will be pushed or pulled by the surrounding charge. The force the test charge experiences (both magnitude and direction), divided by the value of the small test charge, is the electric field vector at that location. Even if you take away the test charge, there is still an electric field at that location.

Electric Field Near an Isolated Point Charge

The electric field around a single isolated point charge, qi is given by,

$$\vec{E} = \frac{1}{4\pi \in_0} \frac{qi}{r^2} \hat{r_i}$$

The electric field direction points straight away from a positive point charge, and straight at a negative point charge. The magnitude of the electric field falls off as $1/$ going away from the point charge.

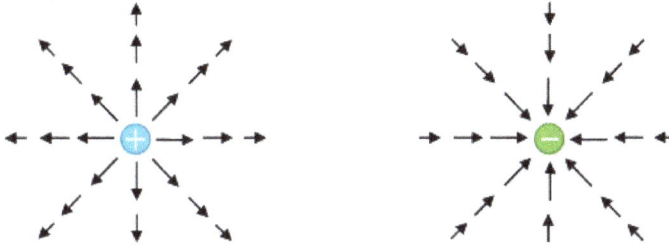

Electric Field Near Multiple Point Charges

If we have multiple charges scattered about, we express the electric field by summing the fields from each individual qi,

$$\vec{E} = \frac{1}{4\pi \in_0} \sum_i \frac{qi}{r^2}\hat{r_i}$$

Where r is the distance between, d_q and the location of interest, while $\hat{r}$ reminds us the direction of the force is in a direct line between d_q and the location of interest.

Types of an Electric Field

The electric field is mainly classified into two types. They are the uniform electric field and the nonuniform electric field.

Uniform Electric Field

When the electric field is constant at every point, then the field is called the uniform electric field. The constant field is obtained by placing the two conductor parallel to each other, and the potential difference between them remains same at every point.

Uniform Electric Field

Non-uniform Electric Field

The field which is irregular at every point is called the non-uniform electric field. The non-uniform field has a different magnitude and directions.

Non-uniform Electric Field

The following are the properties of an electric field.

1. Field lines never intersect each other.

2. They are perpendicular to the surface charge.

3. The field is strong when the lines are close together, and it is weak when the field lines move apart from each other.

4. The number of field lines is directly proportional to the magnitude of the charge.

5. The electric field line starts from the positive charge and ends from negative charge.

6. If the charge is single, then they start or end at infinity.

7. The line curves are continuous in a charge-free region.

When the electric and magnetic field combines, they form the electromagnetic field.

Examples:

1. A line charge:

    Consider a charged line of length L having a uniform linear charge density λ. We will place the line along with x axis with its centre at the origin. We will obtain an expression for the electric field at an arbitrary point p in the xy plane.

    Consider an element of the line of width dx' at the position x'. The distance of the point p(x,y) from the charge element is given by $r'^2 = (x-x')^2 + y^2$ and its position vector with respect to the element is $\vec{r} = \hat{i}(x-x') + \hat{j}y$

    The field at p(x,y) due to the element dx' at

    (x',0)is

    $$d\vec{E} = \frac{1}{4\pi\varepsilon_0} \frac{\lambda dx'}{[(x-x')^2 + y^2]^{3/2}} [(x-x')\hat{i} + y\hat{j}]$$

    The net field at P is

    $$E_x = \frac{\lambda}{4\pi\varepsilon_0} \int_{-L/2}^{L/2} dx' \frac{(x-x')dx'}{[(x-x')^2 + y^2]^{3/2}}$$

    $$E_y = \frac{\lambda}{4\pi\varepsilon_0} \int_{-L/2}^{L/2} dx' \frac{ydx'}{[(x-x')^2 + y^2]^{3/2}}$$

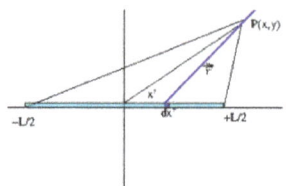

These expressions cannot be evaluated in a closed form for an arbitrary point P. However, if the line charge is taken to be infinitely long, we can evaluate the integrals exactly. In this case, the x component of the field becomes zero by symmetry.

$$E_x = \frac{\lambda}{4\pi\varepsilon_0} \int_{-\infty}^{\infty} \frac{(x-x')dx'}{[(x-x')^2 + y^2]^{3/2}}$$

$$= -\frac{\lambda}{4\pi\varepsilon_0} \int_{-\infty}^{\infty} \frac{zdz}{[z^2 + y^2]^{3/2}}, (z = x - x')$$

$$= 0$$

$$E_y = \frac{\lambda}{4\pi\varepsilon_0} \int_{-\infty}^{\infty} \frac{ydz}{[z^2 + y^2]^{3/2}}$$

(substitute : $z = y\tan\theta$)

$$E_y = \frac{\lambda}{4\pi\varepsilon_0} \int_{-\pi/2}^{\pi/2} \frac{y^2 \sec^2\theta}{y^3 \sec^3\theta} d\theta$$

$$= \frac{\lambda}{4\pi\varepsilon_0 y} \int_{-\pi/2}^{\pi/2} \cos\theta d\theta$$

$$= \frac{\lambda}{2\pi\varepsilon_0 y}$$

2. Field due to a charged ring on its axis

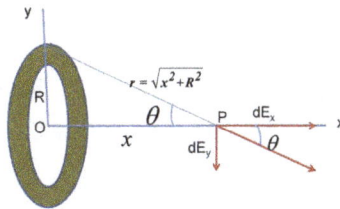

The point P is taken along the x axis at a distance x from the centre of the charged ring. The ring is in the y-z plane. If we take an element on ring, all such elements, by symmetry are located at the same distance $r = \sqrt{x^2 + R^2}$ from the point P. The field at P due to the element is along the radial direction from the element to the point P and can be resolved as shown. By symmetry the y- components would cancel.

$$d\vec{E} = \frac{1}{4\pi\varepsilon} \frac{\lambda dl}{x^2 + R^2} \hat{r}$$

$$dE_x = \frac{1}{4\pi\varepsilon_0} \frac{\lambda dl}{x^2 + R^2} \cos\theta = \frac{1}{4\pi\varepsilon_0} \frac{x\lambda dl}{4\pi\varepsilon_0 (x^2 + R^2)^{3/2}}$$

$$E_x = \frac{\lambda}{4\pi\varepsilon_0} \frac{x}{(x^2 + R^2)^{3/2}} \int_C dl = \frac{Q}{4\pi\varepsilon_0} \frac{x}{(x^2 + R^2)^{3/2}}$$

For X>>R $E_x \Rightarrow \frac{1}{4\pi\varepsilon_0} \frac{Q}{x^2}$

Thus the field is directed along the axis. An interesting point to note is that for large distances, the field has the same form as that due to a point charge. This is to be expected as from such distances, the ring would look like a point.

Electric Potential

Electric potential at a point in an electric field is defined as the amount of work to be done to bring a unit positive electric charge from infinity to that point.

Similarly, the potential difference between two points is defined as the work required to be done for bringing a unit positive charge from one point to other point.

When a body is charged, it can attract an oppositely charged body and can repulse a similar charged body. That means, the charged body has ability of doing work. That ability of doing work of a charged body is defined as electrical potential of that body.

If two electrically charged bodies are connected by a conductor, the electrons starts flowing from lower potential body to higher potential body, that means current starts flowing from higher potential body to lower potential body depending upon the potential difference of the bodies and resistance of the connecting conductor.

So, electric potential of a body is its charged condition which determines whether it will take from or give up electric charge to other body.

Electric potential is graded as electrical level, and difference of two such levels, causes current to flow between them. This level must be measured from a reference zero level. The earth potential is taken as zero level. Electric potential above the earth potential is taken as positive potential and the electric potential below the earth potential is negative.

The unit of electric potential is volt. To bring a unit charge from one point to another, if one joule work is done, then the potential difference between the points is said to be one volt. So, we can say,

$$\text{volt} = \frac{\text{joules}}{\text{coulomb}}$$

If one point has electric potential 5 volt, then we can say to bring one coulomb charge from infinity to that point, 5 joule work has to be done.

If one point has potential 5 volt and another point has potential 8 volt, then 8 – 5 or 3 joules work to be done to move one coulomb from first point to second.

Potential at a Point due to Point Charge

Let us take a positive charge + Q in the space. Let us imagine a point at a distance x from the said charge + Q. Now we place a unit positive charge at that point. As per Coulomb's law, the unit positive charge will experience a force,

$$F = \frac{Q}{4\pi\varepsilon_0\varepsilon_r x^2}$$

Now, let us move this unit positive charge, by a small distance dx towards charge Q.

During this movement the work done against the field is,

$$dw = -F.dx = -\frac{Q}{4\pi\in_0\in_r x^2}$$

So, total work to be done for bringing the positive unit charge from infinity to distance x, is given by,

$$-\int_\infty^x dw = -\int_\infty^x \frac{Q}{4\pi\in_0\in_r x^2}\cdot dx = \frac{Q}{4\pi\in_0\in_r}\left[\frac{1}{x}\right]_\infty^x = \frac{Q}{4\pi\in_0\in_r}\left[\frac{1}{x}-\frac{1}{\infty}\right] = \frac{Q}{4\pi\in_0\in_r x}$$

As per definition, this is the electric potential of the point due to charge + Q. So, we can write,

$$V = \frac{Q}{4\pi\in_0\in_r x}$$

Potential Difference between Two Points

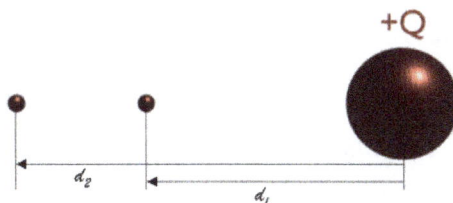

Let us consider two points at distance d_1 meter and d_2 meter from a charge +Q.

We can express the electric potential at the point d1 meter away from +Q, as,

$$V_{d1} = \frac{Q}{4\pi \in_0 \in_r d_1}$$

We can express the electric potential at the point d2 meter away from +Q, as,

$$V_{d2} = \frac{Q}{4\pi \in_0 \in_r d_2}$$

Thus, the potential difference between these two points is

$$V_{d1} - V_{d2} = \frac{Q}{4\pi \in_0 \in_r d_1} - \frac{Q}{4\pi \in_0 \in_r d_2} = \frac{Q}{4\pi \in_0 \in_r}\left[\frac{1}{d_1} - \frac{1}{d_2}\right]$$

Electric Flux

Electric flux is the rate of flow of the electric field through a given area. Electric flux is proportional to the number of electric field lines going through a virtual surface.

If the electric field is uniform, the electric flux passing through a surface of vector area S is $\phi_E = E.S = ES\cos\theta$ where E is the magnitude of the electric field (having units of V/m), S is the area of the surface, and θ is the angle between the electric field lines and the normal (perpendicular) to S.

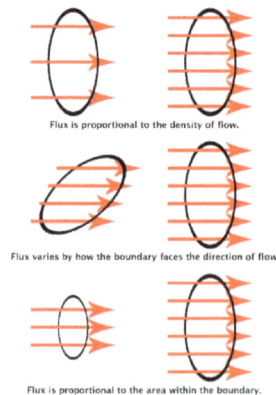

Flux is proportional to the density of flow.

Flux varies by how the boundary faces the direction of flow.

Flux is proportional to the area within the boundary.

For a non-uniform electric field, the electric flux dΦE through a small surface area dS is given by $d\Phi_E = E \cdot dS$ (the electric field, E, multiplied by the component of area perpendicular to the field).

Gauss' Law describes the electric flux over a surface S as the surface integral: $\Phi_E = \iint_S E \cdot dS$

where E is the electric field and dS is a differential area on the closed surface S with an outward facing surface normal defining its direction.

It is important to note that while the electric flux is not affected by charges that are not within the closed surface, the net electric field, E, in the Gauss' Law equation, can be affected by charges that lie outside the closed surface. While Gauss' Law holds for all situations, it is only useful for "by hand" calculations when high degrees of symmetry exist in the electric field. Examples include spherical and cylindrical symmetry.

Electric flux has SI units of volt metres (V m), or, equivalently, newton metres squared per coulomb (N m^2 C^{-1}). Thus, the SI base units of electric flux are kg·m^3·s^{-3}·A^{-1}.

Gauss's Law

Gauss Law states that the total electric flux out of a closed surface is equal to the charge enclosed divided by the permittivity. The electric flux in an area is defined as the electric field multiplied by the area of the surface projected in a plane and perpendicular to the field.

According to the Gauss law, the total flux linked with a closed surface is $1/\varepsilon 0$ times the charge enclosed by the closed surface.

$$\oint \vec{E}.\overrightarrow{ds} = \frac{1}{\epsilon_0}q.$$

For example, A point charge q is placed inside a cube of edge 'a'. Now as per the Gauss law, the flux through each face of the cube is $q/6\varepsilon_0$.

The electric field is the basic concept to know about electricity. Generally, the electric field of the surface is calculated by applying Coulomb's law, but to calculate the electric field distribution in a closed surface, we need to understand the concept of Gauss law. It explains about the electric charge enclosed in a closed or the electric charge present in the enclosed closed surface.

Gauss Law Formula

As per the Gauss theorem, the total charge enclosed in a closed surface is proportional to the total flux enclosed by the surface. Therefore, If ϕ is total flux and $\epsilon 0$ is electric constant, the total electric charge Q enclosed by the surface is;

$$Q = \phi \, \epsilon_0$$

The Gauss law formula is expressed by;

$$\phi = Q/ \epsilon_0$$

Where,

Q = total charge within the given surface,

ε_0 =the electric constant.

The Gauss Theorem

The net flux through a closed surface is directly proportional to the net charge in the volume enclosed by the closed surface.

$$\Phi = \rightarrow E.d \rightarrow A = q_{net} / \varepsilon_0$$

In simple words, the Gauss theorem relates the 'flow' of electric field lines (flux) to the charges within the enclosed surface. If there are no charges enclosed by a surface, then the net electric flux remains zero.

This means that the number of electric field lines entering the surface is equal to the field lines leaving the surface.

The Gauss theorem statement also gives an important corollary:

The electric flux from any closed surface is only due to the sources (positive charges) and sinks (negative charges) of electric fields enclosed by the surface. Any charges outside the surface do not contribute to the electric flux. Also, only electric charges can act as sources or sinks of electric fields. Changing magnetic fields, for example, cannot act as sources or sinks of electric fields.

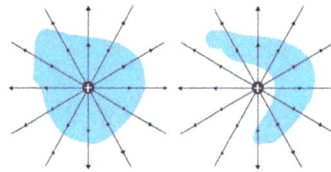

The Gauss Theorem

he net flux for the surface on the left is non-zero as it encloses a net charge. The net flux for the surface on the right is zero since it does not enclose any charge.

⇒ Note: The Gauss law is only a restatement of the Coulombs law. If you apply the Gauss theorem to a point charge enclosed by a sphere, you will get back the Coulomb's law easily.

Applications of Gauss Law

1. In the case of a charged ring of radius R on its axis at a distance x from the centre of the ring. $E = \dfrac{1}{4\pi \in_0} \dfrac{qx}{\left(R^2 + x^2 \right)^{3/2}}$. At the centre, x = 0 and E = 0.

2. In case of an infinite line of charge, at a distance 'r'. E = $(1/4 \times \pi r \varepsilon_0)$ $(2\pi/r)$ = $\lambda/2\pi r \varepsilon_0$. Where λ is the linear charge density.

3. The intensity of the electric field near a plane sheet of charge is E = $\sigma/2\varepsilon_0 K$ where σ = surface charge density.

4. The intensity of the electric field near a plane charged conductor E = $\sigma/K\varepsilon 0$ in a medium of dielectric constant K. If the dielectric medium is air, then Eair = σ/ε_0.

5. The field between two parallel plates of a condenser is E = $\sigma/\varepsilon 0$, where σ is the surface charge density.

Electric Field due to Infinite Wire – Gauss Law Application

Consider an infinitely long line of charge with the charge per unit length being λ. We can take advantage of the cylindrical symmetry of this situation. By symmetry, The electric fields all point radially away from the line of charge, There is no component parallel to the line of charge.

We can use a cylinder (with an arbitrary radius (r) and length (l)) centred on the line of charge as our Gaussian surface.

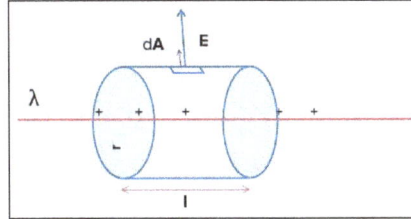

Applications of Gauss Law – Electric Field due to Infinite Wire

As you can see in the above diagram, the electric field is perpendicular to the curved surface of the cylinder. Thus, the angle between the electric field and area vector is zero and $\cos \theta = 1$

The top and bottom surfaces of the cylinder lie parallel to the electric field. Thus the angle between area vector and the electric field is 90 degrees and $\cos \theta = 0$.

Thus, the electric flux is only due to the curved surface.

According to Gauss Law,

$$\Phi = E.d \rightarrow A$$

$$\Phi = \Phi_{curved} + \Phi_{top} + \Phi_{bottom}$$

$$\Phi = \rightarrow E.d \rightarrow A = \int E.dA \cos 0 + \int E.dA \cos 90^\circ + \int E.dA \cos 90^\circ$$

$$\Phi = \int E.dA \times 1$$

Due to radial symmetry, the curved surface is equidistant from the line of charge and the electric field in the surface has a constant magnitude throughout.

$$\Phi = \int E.dA = E \int dA = E.2\pi rl$$

The net charge enclosed by the surface is:

$$q_{net} = \lambda.l$$

Using Gauss theorem,

$$\Phi = E \times 2\pi rl = q_{net} / \varepsilon_{0=} \lambda l / \varepsilon_0$$

$$E \times 2\pi rl = \lambda l / \varepsilon_0$$

$$E = \lambda l / 2\pi r \varepsilon_0$$

Problems on Gauss Law

Example: A uniform electric field of magnitude E = 100 N/C exists in the space in X-direction. Using the Gauss theorem calculate the flux of this field through a plane square area of edge 10 cm placed in the Y-Z plane. Take the normal along the positive X-axis to be positive.

Solution:

The flux $\Phi = \int E.\cos\theta ds$.

As the normal to the area points along the electric field, $\theta = 0..$

Also, E is uniform so, $\Phi = E.\Delta S = (100\,N/C)(0.10m)^2 = 1\,N-m^2$

Example: A large plane charge sheet having surface charge density σ = 2.0 × 10⁻⁶ C-m⁻² lies in the X-Y plane. Find the flux of the electric field through a circular area of radius 1 cm lying completely in the region where x, y, z are all positive and with its normal making an angle of 60° with the Z-axis.

Solution:

The electric field near the plane charge sheet is $E = \sigma/2\varepsilon_0$ in the direction away from the sheet. At the given area, the field is along the Z-axis.

The area $= \pi r^2 = 3.14 \times 1 cm^2 = 3.14 \times 10^{-4} m^2$

The angle between the normal to the area and the field is $60°$

Hence, according to Gauss theorem, the flux $= \overline{E}\cdot\overline{\Delta S} = E.\Delta S\cos\theta = \sigma/2\varepsilon_0 \times pr^2 \cos 60°$

$$= \frac{2.0\times10^{-6}C/m^2}{2\times 8.85\times10^{-12}C^2/N-m^2}\times\left(3.14\times10^{-4}m^2\right)\frac{1}{2} = 17.5\,N-m^2c^{-1}$$

Example: A charge of 4×10⁻⁸ C is distributed uniformly on the surface of a sphere of radius 1 cm. It is covered by a concentric, hollow conducting sphere of radius 5 cm.

   •   Find the electric field at a point 2 cm away from the centre.

   •   A charge of 6 × 10⁻⁸C is placed on the hollow sphere. Find the surface charge density on the outer surface of the hollow sphere.

Solution:

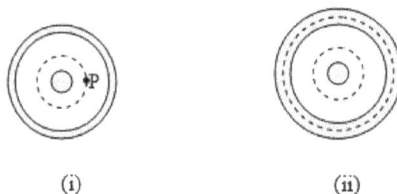

(i)   (ii)

(a) Let us consider the figure above (i).

Suppose, we have to find the field at point P. Draw a concentric spherical surface through P. All the points on this surface are equivalent and by symmetry, the field at all these points will be equal in magnitude and radial in direction.

The flux through this surface $= \oint \vec{E}.\overrightarrow{dS}$

$$= \oint EdS = E\oint dS = 4\pi x^2 E.$$

where $x = 2$ cm $= 2 \times 10^{-2}$ m.

From Gauss law, this flux is equal to the charge q contained inside the surface divided by ε_0. Thus,

$$\Rightarrow 4\pi x^2 E = q / \varepsilon_0 \ or, \ E = q / 4\pi\varepsilon_0 x^2$$
$$= (9\times10^9)\times\left[(4\times10^{-8})/(4\times10^{-4})\right] = 9\times10^5 \ NC^{-1}$$

(b) Let us consider the figure above (ii).

Take the Gaussian surface through the material of the hollow sphere. As the electric field in a conducting material is zero, the flux $\oint \vec{E}.\overrightarrow{dS}$ through this Gaussian surface is zero.

Using Gauss law, the total charge enclosed must be zero. Hence, the charge on the innersurface of the hollow sphere is 4×10^{-8}C.

But the total charge given to this hollow sphere is 6×10^{-8} C. Hence, the charge on the outer surface will be 10×10^{-8}C.

Example: The figure shows three concentric thin spherical shells A, B and C of radii a, b, and c respectively. The shells A and C are given charges q and -q respectively and the shell B is earthed. Find the charges appearing on the surfaces of B and C.

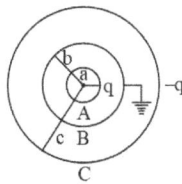

Solution:

The inner surface of B must have a charge -q from the Gauss law. Suppose, the outer surface of B has a charge q'.

The inner surface of C must have a charge -q' from Gauss law. As the net charge on C must be -q, its outer surface should have a charge q' – q. The charge distribution is shown in the figure.

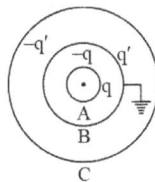

The potential at B,

- Due to the charge q on A = $q/4\pi\varepsilon_0 b$,

- Due to the charge -q on the inner surface of B = $-q/4\pi\varepsilon_0 b$,

- Due to the charge q' on the outer surface of B = $q'/4\pi\varepsilon_0 b$,

- Due to the charge -q', on the inner surface of C = $-q'/4\pi\varepsilon_0 c$,

- Due to the charge q' − q on the outer surface of C = $(q' - q)/4\pi\varepsilon_0 c$.

The net potential is, VB = $q'/4\pi\varepsilon_0 b - q/4\pi\varepsilon_0 c$

This should be zero as the shell B is earthed. Thus, q' = q × b/c

The charges on various surfaces are as shown in the figure:

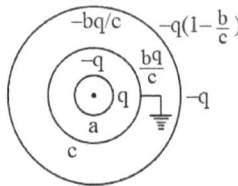

Example: A particle of mass 5 × 10⁻⁶g is kept over a large horizontal sheet of charge of density 4.0 × 10⁻⁶ C/m². What charge should be given to this particle so that if released, it does not fall down? How many electrons are to be removed to give this charge? How much mass is decreased due to the removal of these electrons?

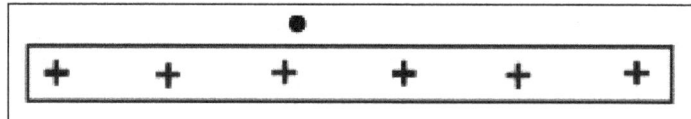

Solution:

The electric field in front of the sheet is,

$$E - \sigma/2\varepsilon_0 = \left(4.0\times10^{-6}\right)/\left(2\times8.85\times10^{-12}\right)2.26\times10^5 N/C$$

If a charge q is given to the particle, the electric force qE acts in the upward direction. It will balance the weight of the particle if

$$q\times2.26\times10^5 N/C = 5\times10^{-9} kg\times9.8 m/s^2$$

Or, $$q = \left[4.9\times10^{-8}\right]/\left[2.26\times10^5\right]C = 2.21\times10^{-13}C$$

The charge on one electron is 1.6 × 10⁻¹⁹C. The number of electrons to be removed;

$$= \left[2.21\times10^{-13}\right]/\left[1.6\times10^{-19}\right] = 1.4\times10^6$$

Mass decreased due to the removal of these electrons $= 1.4 \times 10^6 \times 9.1 \times 10^{-31} kg = 1.3 \times 10^{-24} kg.$

Example: Two conducting plates A and B are placed parallel to each other. A is given a charge Q_1 and B a charge Q_2. Find the distribution of charges on the four surfaces.

Solution:

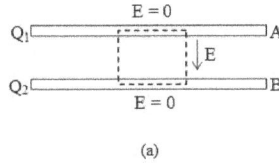

(a)

Consider a Gaussian surface as shown in figure (a). Two faces of this closed surface lie completely inside the conductor where the electric field is zero.

The flux through these faces is, therefore, zero. The other parts of the closed surface which are outside the conductor are parallel to the electric field and hence the flux on these parts is also zero.

The total flux of the electric field through the closed surface is, therefore, zero. From Gauss law, the total charge inside the closed surface should be zero. The charge on the inner surface of A should be equal and opposite to that on the inner surface of B.

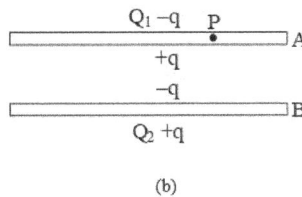

(b)

The distribution should be like the one shown in figure (b). To find the value of q, consider the field at a point P inside the plate A. Suppose, the surface area of the plate (one side) is A.

Using the equation $E = \sigma/2\varepsilon_0$, the electric field at P;

- Due to the charge $Q_1 - q = (Q_1 - q)/2A\varepsilon_0$ (downward),
- Due to the charge $+q = q/2A\varepsilon_0$ (upward),
- Due to the charge $-q = q/2A\varepsilon_0$ (downward),
- Due to the charge $Q_2 + q = (Q_2 + q)/2A\varepsilon_0$ (upward).

The net electric field at P due to all the four charged surfaces is (in the downward direction)

$$(Q_1 - q)/2A\varepsilon_0 - q/2A\varepsilon_0 + q/2A\varepsilon_0 - (Q_2 + q)/2A\varepsilon_0$$

As the point P is inside the conductor, this field is should be zero.

Hence, $Q_1 - q - Q_2 - q = 0$

Or $q = (Q_1 - Q_2)/2$

Thus, $Q_1 - q = (Q_1 + Q_2)/2$

and $Q_2 + q = [Q_1 + Q_2]/2$

Using these equations, the distribution shown in the figures can be redrawn as in the figure.

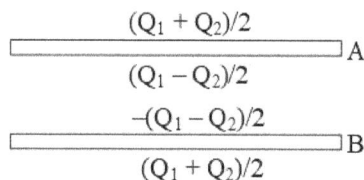

When charged conducting plates are placed parallel to each other, the two outermost surfaces get equal charges and the facing surfaces get equal and opposite charges.

Example: A solid conducting sphere having a charge Q is surrounded by an uncharged concentric conducting hollow spherical shell. Let the potential difference between the surface of the solid sphere and that of the outer surface of hollow shell be V. What will be the new potential difference between the same two surfaces if the shell is given a charge -3Q?

Solution:

In case of a charged conducting sphere

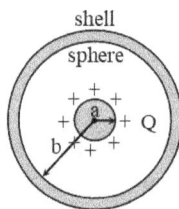

$$V_{in} = V_c = V_s = 1/4\pi\varepsilon_0$$

and $V_{out} = 1/4\pi\varepsilon_0$

So if a and b are the radii of a sphere and spherical shell respectively,

the potential at their surfaces will be;

Vsphere $= 1/4\pi\varepsilon_0 [Q/a]$ and Vshell $= 1/4\pi\varepsilon_0 [Q/b]$ and so according to the given problem;

$$V = V'sphere - V'shell = Q/4\pi\varepsilon_0 [1/a - 1/b] = V$$

Now when the shell is given a charge (-3Q) the potential at its surface and also inside will change by;

$$V_0 = 1/4\pi\varepsilon_0 [-3Q/b]$$

So that now,

V'sphere $1/4\pi\varepsilon_0\left[Q/a+V_0\right]$ and V'shell $=1/4\pi\varepsilon_0\left[Q/b+V_0\right]$

Hence, V'sphere $-$ V'shell $=Q/4\pi\varepsilon_0\left[1/a-1/b\right]=V$

i.e., if any charge is given to external shell the potential difference between sphere and shell will not change.

This is because by the presence of charge on the outer shell, potential everywhere inside and on the surface of the shell will change by the same amount and hence the potential difference between sphere and shell will remain unchanged.

Example: A very small sphere of mass 80 g having a charge q is held at height 9 m vertically above the centre of a fixed non conducting sphere of radius 1 m, carrying an equal charge q. When released it falls until it is repelled just before it comes in contact with the sphere. Calculate the charge q. [g = 9.8 m/s²]

Solution:

Keeping in mind that here both electric and gravitational potential energy is changing and for an external point, a charged sphere behaves as the whole of its charge were concentrated at its centre.

Applying the law of conservation of energy between initial and final position, we have

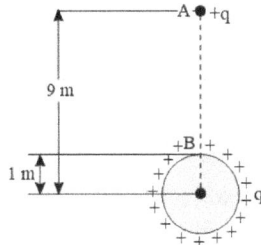

$$=1/4\pi\varepsilon_0\times\left(q.q/9\right)+mg\times9=1/4\pi\varepsilon_0\times\left(q^2/1\right)+mg\times1$$
$$or, q_2=\left(80\times10^{-3}\times9.8\right)/10^9=28\mu C$$

Magnetic Force

The magnetic force is a consequence of the electromagnetic force, one of the four fundamental forces of nature, and is caused by the motion of charges. Two objects containing charge with the same direction of motion have a magnetic attraction force between them. Similarly, objects with charge moving in opposite directions have a repulsive force between them.

Consider two objects. The magnitude of the magnetic force between them depends on how much charge is in how much motion in each of the two objects and how far apart they are. The direction of the force depends on the relative directions of motion of the charge in each case.

The usual way to go about finding the magnetic force is framed in terms of a fixed amount of charge q moving at constant velocity v in a uniform magnetic field B. If we don't know the magnitude of the magnetic field directly then we can still use this method because it is often possible to calculate the magnetic field based on the distance to a known current.

The magnetic force is described by the Lorentz Force law:

$$\vec{F} = q\vec{v} \times \vec{B}$$

In this form it is written using the vector cross product. We can write the magnitude of the magnetic force by expanding the cross product. Written in terms of the angle $\theta \left(< 180° \right)$ between the velocity vector and the magnetic field vector:

$$F = qvB\sin\theta$$

The direction of the force can be found using the right-hand-slap rule. This rule describes the direction of the force as the direction of a 'slap' of an open hand. As with the right-hand-grip rule, the fingers point in the direction of the magnetic field. The thumb points in the direction that positive charge is moving. If the moving charge is negative (for example, electrons) then you need to reverse the direction of your thumb because the force will be in the opposite direction. Alternatively, you can use your left hand for moving negative charge.

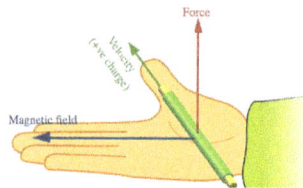

Figure: Using the right-hand-slap rule for the force due to a positive charge moving in a magnetic field.

Sometimes we want to find the force on a wire carrying a current I in a magnetic field. This can be done by rearranging our previous expression. If we recall that velocity is a distance / time then if a wire has length L we can write =

$$qv = \frac{qL}{t}$$

and since current is the amount of charge flowing per second,

$$qv = IL$$

and therefore

$$F = BIL\sin\theta$$

Magnetic Field

A magnetic field is generated when electric charge carriers such as electrons move through space or within an electrical conductor. The geometric shapes of the magnetic flux lines produced by moving charge carriers (electric current) are similar to the shapes of the flux lines in an electrostatic field. But there are differences in the ways electrostatic and magnetic fields interact with the environment.

Electrostatic flux is impeded or blocked by metallic objects. Magnetic flux passes through most metals with little or no effect, with certain exceptions, notably iron and nickel. These two metals, and alloys and mixtures containing them, are known as ferromagnetic materials because they concentrate magnetic lines of flux. An electromagnet provides a good example. An air-core coil carrying direct current produces a magnetic field. If an iron core is substituted for the air core in a given coil, the intensity of the magnetic field is greatly increased in the immediate vicinity of the coil. If the coil has many turns and carries a large current, and if the core material has exceptional ferromagnetic properties, the flux density near the ends of the core (the poles of the magnet) can be such that the electromagnet can be used to pick up and move cars.

When charge carriers are accelerated (as opposed to moving at constant velocity), a fluctuating magnetic field is produced. This generates a fluctuating electric field, which in turn produces another varying magnetic field. The result is a "leapfrog" effect, in which both fields can propagate over vast distances through space. Such a synergistic field is known as an electromagnetic field. This is the phenomenon that makes wireless communications and broadcasting possible.

Magnetic Lines of Force

It can be defined as curved lines used to represent a magnetic field, drawn such that the number of lines relates to the magnetic field's strength at a given point and the tangent of any curve at a particular point is along the direction of magnetic force at that point.

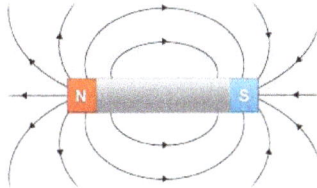

Properties:

1. Magnetic lines of force start from the North Pole and end at the South Pole.

2. They are continuous through the body of a magnet.

3. Magnetic lines of force can pass through iron more easily than air.

4. Two magnetic lines of force can not intersect each other.

5. They tend to contract longitudinally.

6. They tend to expand laterally.

Biot-Savart Law

In Physics, Biot-Savart law is a fundamental quantitative relationship between an electric current and the magnetic field.

Let a certain conductor be carrying a current 'I' in a direction on in the figure. Let 'P' be the point where the magnetic field due to the wire is to be studied. Let a small portion be considered which is of length 'dl'. Let the line joining 'dl' and point 'P' from an angle θ with the tangent to 'dl'.

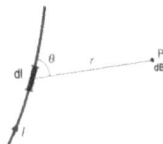

Since the portion considered is very small, the magnetic field given by it at point P will also be small. By experimental observations and empirically also, dB is found to depend on several factors.

i)   Here dB is the measurement of magnetic energy which arises from the electrical energy represented by I which act respectively as output and input. Therefore they should have direct dependency. i.e

$$dB \propto I$$

ii) In a certain length of a conductor, certain amount of charge is present at a moment and the magnetic effect it can produce depend on the total number of charges, which in turn depend on the length consider, i.e.

$$dB \propto dl$$

iii) Any force or phenomena which spread out spherically have inverse proportionality to the square of the distance between the source and point of observation.

$$dB \propto \frac{1}{r^2}$$

iv) Similarly the magnetic field is found to be least when the angle between r and dl is the smallest ($0°$) and it is largest when the angle is $90°$. So,

$$dB \propto \sin\theta$$

Therefore overall,

$$dB \propto \frac{Idl\sin\theta}{r^2}$$

Or, $dB = k\dfrac{Idl\sin\theta}{r^2}$

In SI units, the value of k $= \dfrac{\mu_0}{4\pi} = 10^{-7} Hm^{-1}$

Or, $dB = \dfrac{\mu_0}{4\pi}\dfrac{Idl\sin\theta}{r^2}$

This expression is called as Biot Savart Law or Laplace Law. It is the basic formula to find the magnetic field due to any structures for which all the dB's along the length of the structure have to be added to find out the total B.

Ampere's Law

Ampere's law allows us to calculate magnetic fields from the relation between the electric currents that generate this magnetic fields. It states that for a closed path the sum over elements of the component of the magnetic field is equal to electric current multiplied by the empty's permeability.

Integration over the closed path of (magnetic field . infenitesimal segment of the integration path) = empty's permeability * enclosed electric current by the path.

The equation is:

$$\int B.dl = \mu_0 I$$

Where:

B : magnetic field

dl : infinitesimal segment of the integration path

μ_0 : empty's permeability

I : enclosed electric current by the path

Ampere's Law Formula Examples

1) Determine the magnetic field strength a distance r away from an infinitely long current carrying wire using the Ampere's law.

Answer:

From the Ampere's law, we solve the integral

$$\int B.dl = B\int dl = B2\pi r$$

Then,

$$B2\pi r = \mu_0 I$$
$$B = \mu_0 I / 2\pi r$$

2) Determine the magnetic field strength anywhere inside a solenoid with n turns per unit length using the Ampere's law

Answer:

Solving the integral we have

$$\int B.dl = B\int dl = Bl$$

Now the current inside the solenoid is

$I = ni$, where n is number of turns

We define N as $N = n / l$

So, $B = \mu_o ni / l$

Finally, $B = \mu_o Ni$

Faraday's Law

Faraday's law of electromagnetic induction which is also known as Faraday's law is the basic law of electromagnetism which explains the working principle of motors, generators, inductors, and electrical transformers. It helps to understand key points leading to the practical generation of electricity or electromagnetic induction.

The law was proposed in the year 1831 by an experimental physicist and chemist named Michael Faraday. So you can see where the name of the law comes from. That being said, the Faraday's law or laws of electromagnetic induction are basically the results or the observations of the experiments that Faraday conducted. He performed three main experiments to discover the phenomenon of electromagnetic induction.

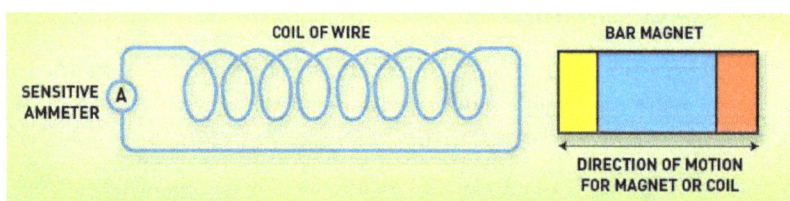

Faraday's First Law

Faraday's first law of electromagnetic induction states that whenever a conductor is placed in the varying magnetic field, electromagnetic fields are induced known as induced emf. If the conductor circuit is closed, a current is also induced which are called induced current.

Ways of Changing the Magnetic Field

- By rotating the coil relative to the magnet.

- By moving the coil into or out of the magnetic field.

- By changing the area of a coil placed in the magnetic field.
- By moving a magnet towards or away from the coil.

Faraday's Second Law

Faraday's second law of electromagnetic induction states that the induced emf in a coil is equal to the rate of change of flux linkage. Here the flux is nothing but the product of the number of turns in the coil and flux connected with the coil.

Faraday's Law Formula

The formula of Faraday's law is given below:

$$\varepsilon = -N\frac{\Delta\phi}{\Delta t}$$

Where,

- ε is the electromotive force
- Φ is the magnetic flux
- N is the number of turns

The negative sign indicates that the direction of the induced emf and change in direction of magnetic fields have opposite signs.

Faraday's Experiment

Relationship Between Induced EMF and Flux

In the first experiment, he proved that when the strength of the magnetic field is varied then only the induced current is produced. An ammeter was connected to a loop of wire; the ammeter deflected when a magnet was moved towards the wire.

In the second experiment, he proved that passing a current through an iron rod would make it electromagnetic. He observed that when there a relative motion exists between the magnet and the coil, an induced electromagnetic force is created. When the magnet was rotated about its axis, no electromotive force was observed but when the magnet was rotated about its own axis then the induced electromotive force was produced. Thus, there was no deflection in the ammeter when the magnet was held stationary.

While conducting the third experiment, he recorded that galvanometer did not show any deflection and no induced current was produced in the coil when the coil was moved in a stationary magnetic field. The ammeter deflected in the opposite direction when the magnet was moved away from the loop.

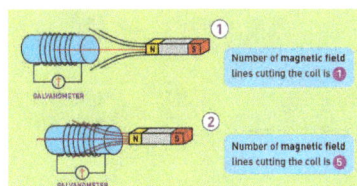

Position of Magnet	Deflection in Galvanometer
Magnet at Rest	No deflection in the galvanometer
The magnet moves towards the coil	Deflection in the galvanometer in one direction
Magnet is held stationary at the same position (near the coil)	No deflection galvanometer
The magnet moves away from the coil	Deflection in galvanometer but in the opposite direction
The magnet held stationary at the same position (away from the coil)	No deflection in the galvanometer

After conducting all the experiments, Faraday finally concluded that if relative motion existed between a conductor and a magnetic field, the flux linkage with a coil changed and this change in flux produced a voltage across a coil.

Faraday law basically states, "when the magnetic flux or the magnetic field changes with time, the electromotive force is produced". Additionally, Michael Faraday also formulated two laws on the basis of the above experiments.

Applications of Faraday's Law

Following are the fields where Faraday's law find applications:

- Electrical equipment like transformers works on the basis of Faraday's law.

- Induction cooker works on the basis of mutual induction which is the principle of Faraday's law.

- By inducing an electromotive force into an electromagnetic flowmeter, the velocity of the fluids is recorded.

Lenz's Law

Lenz's law of electromagnetic induction states that the direction of the current induced in a conductor by a changing magnetic field is such that the magnetic field created by the induced current opposes the initial changing magnetic field which produced it. The direction of this current flow is given by Fleming's right hand rule.

This can be hard to understand at first – so let's look at an example problem. Remember that when a current is induced by a magnetic field, the magnetic field that this induced current produces will create its own magnetic field. This magnetic field will always be such that it opposes the magnetic field that originally created it. In the example below, if the magnetic field "B" is increasing – as shown in (1) – the induced magnetic field will act in opposition to it.

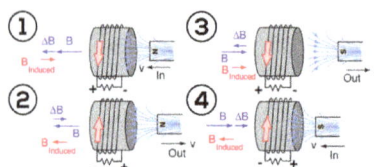

When the magnetic field "B" is decreasing – as shown in the figure above – the induced magnetic field will again act in opposition to it. But this time 'in opposition' means that it is acting to increase the field – since it is opposing the decreasing rate of change.

Lenz's law is based on Faraday's law of induction. Faraday's law tells us that a changing magnetic field will induce a current in a conductor. Lenzs law tells us the direction of this induced current, which opposes the initial changing magnetic field which produced it. This is signified in the formula for Faraday's law by the negative sign ('–').

$$\varepsilon = -\frac{d\Phi_B}{dt}$$

This change in the magnetic field may be caused by changing the magnetic field strength by moving a magnet towards or away from the coil, or moving the coil into or out of the magnetic field. In other words, we can say that the magnitude of the EMF induced in the circuit is proportional to the rate of change of flux.

$$\xi \alpha \frac{d\Phi}{dt}$$

Lenz's Law Formula

Lenz's law states that when an EMF is generated by a change in magnetic flux according to Faraday's Law, the polarity of the induced EMF is such, that it produces an induced current whose magnetic field opposes the initial changing magnetic field which produced it.

The negative sign used in Faraday's law of electromagnetic induction, indicates that the induced EMF (ε) and the change in magnetic flux ($\delta\Phi_B$) have opposite signs. The formula for Lenz's law is shown below:

$$\varepsilon = -N\frac{\partial\Phi_B}{\partial t}$$

Where:

- ε = Induced emf
- $\delta\Phi_B$ = change in magnetic flux
- N = No of turns in coil

Lenz's Law and Conservation of Energy

To obey the conservation of energy, the direction of the current induced via Lenz's law must create a magnetic field that opposes the magnetic field that created it. In fact, Lenz's law is a consequence of the law of conservation of energy.

If the magnetic field created by the induced current is the same direction as the field that produced it, then these two magnetic fields would combine and create a larger magnetic field. This combined larger magnetic field would, in turn, induce another current within the conductor twice the magnitude of the original induced current.

And this would, in turn, create another magnetic field which would induce yet another current. And so on. So we can see that if Lenz's law did not dictate that the induced current must create a magnetic field that opposes the field that created it – then we would end up with an endless positive feedback loop, breaking the conservation of energy (since we are effectively creating an endless energy source).

Lenz's law also obeys Newton's third law of motion (i.e to every action there is always an equal and opposite reaction). If the induced current creates a magnetic field which is equal and opposite to the direction of the magnetic field that creates it, then only it can resist the change in the magnetic field in the area. This is in accordance with Newton's third law of motion.

To better understand Lenz's law, let us consider two cases:

Case 1: When a magnet is moving towards the coil.

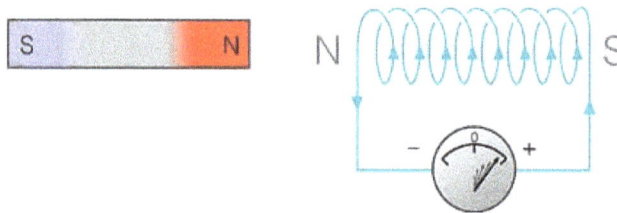

When the north pole of the magnet is approaching towards the coil, the magnetic flux linking to the coil increases. According to Faraday's law of electromagnetic induction, when there is a change in flux, an EMF and hence current is induced in the coil and this current will create its own magnetic field.

Now according to Lenz's law, this magnetic field created will oppose its own or we can say opposes the increase in flux through the coil and this is possible only if approaching coil side attains north polarity, as we know similar poles repel each other. Once we know the magnetic polarity of the coil side, we can easily determine the direction of the induced current by applying right hand rule. In this case, the current flows in the anticlockwise direction.

Case 2: When a magnet is moving away from the coil.

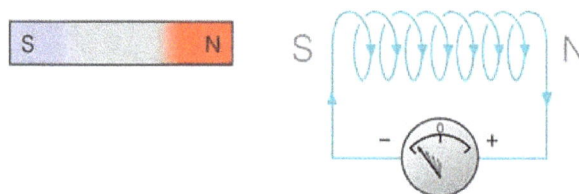

When the north pole of the magnet is moving away from the coil, the magnetic flux linking to the coil decreases. According to Faraday's law of electromagnetic induction, an EMF and hence current is induced in the coil and this current will create its own magnetic field.

Now according to Lenz's law, this magnetic field created will oppose its own or we can say opposes the decrease in flux through the coil and this is possible only if approaching coil side attains south polarity, as we know dissimilar poles attract each other. Once we know the magnetic polarity of the coil side, we can easily determine the direction of the induced current by applying right hand rule. In this case, the current flows in a clockwise direction.

Note that for finding the directions of magnetic field or current, use the right-hand thumb rule i.e if the fingers of the right hand are placed around the wire so that the thumb points in the direction of current flow, then the curling of fingers will show the direction of the magnetic field produced by the wire.

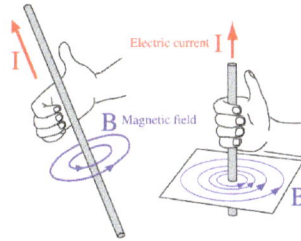

Lenzs law can be stated as follows:

*   If the magnetic flux Φ linking a coil increases, the direction of current in the coil will be such that it will oppose the increase in flux and hence the induced current will produce its flux in a direction as shown below (using Fleming's right-hand thumb rule).

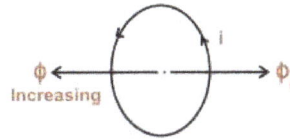

*   If magnetic flux Φ linking a coil is decreasing, the flux produced by the current in the coil is such, that it will aid the main flux and hence the direction of current is as shown below.

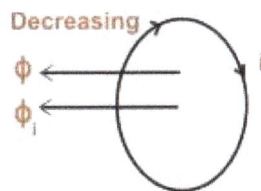

Maxwell's Equations

Maxwell's equations are a set of four differential equations that form the theoretical basis for describing classical electromagnetism:

*   Gauss' law: Electric charges produce an electric field. The electric flux across a closed surface is proportional to the charge enclosed.

*   Gauss' law for magnetism: There are no magnetic monopoles. The magnetic flux-and-faradays-law-quantitative across a closed surface is zero.

- Faraday's law: Time-varying magnetic fields produce an electric field.

- Ampère's law: Steady currents and time-varying electric fields (the latter due to Maxwell's correction) produce a magnetic field.

First assembled together by James Clerk 'Jimmy' Maxwell in the 1860s, Maxwell's equations specify the electric and magnetic fields and their time evolution for a given configuration.

Integral Form

Where q and v are respectively the electric charge and velocity of a particle, the Lorentz law defines the electric field $\mathbf{E}$ and magnetic field $\mathbf{B}$ by specifying the total electromagnetic force $\mathbf{F}$ as

$$\mathbf{F} = q\mathbf{E} + qv \times \mathbf{B}.$$

In essence, one takes the part of the electromagnetic force that arises from interaction with moving charge qv as the magnetic field and the other part to be the electric field.

Gauss' law. The earliest of the four Maxwell's equations to have been discovered (in the equivalent form of Coulomb's law) was Gauss' law. In its integral form in SI units, it states that the total charge contained within a closed surface is proportional to the total electric flux (sum of the normal component of the field) across the surface:

$$\int_s \mathbf{E} \cdot d\mathbf{a} = \frac{1}{\epsilon_0} \int p \, dV,$$

where the constant of proportionality is $1/\epsilon_0$, the reciprocal of the electric constant. The total charge is expressed as the charge density ρ integrated over a region.

Gauss' law for magnetism. Although magnetic dipoles can produce an analogous magnetic flux, which carries a similar mathematical form, there exist no equivalent magnetic monopoles, and therefore the total "magnetic charge" over all space must sum to zero. Therefore, Gauss' law for magnetism reads simply:

$$\int_s \mathbf{B} \cdot d\mathbf{a} = 0.$$

Faraday's law. The electric and magnetic fields become intertwined when the fields undergo time evolution. In the 1820s, Faraday discovered that a change in magnetic flux produces an electric field over a closed loop. This relation is now called Faraday's law:

$$\int_{loop} \mathbf{E} \cdot d\mathbf{s} = \frac{d}{dt} \int_s \mathbf{B} \cdot d\mathbf{a}.$$

With the orientation of the loop defined according to the right hand rule, the negative sign reflects Lenz's law.

Ampère's law. Finally, Ampère's law suggests that steady current across a surface leads to a magnetic field (expressed in terms of flux). In addition, Maxwell determined that that rapid changes in the electric flux $(d/dt)\mathbf{E}\cdot d\mathbf{a}$ can also lead to changes in magnetic flux. Altogether, Ampère's law with Maxwell's correction holds that

$$\int_{loop}\mathbf{B}\cdot d\mathbf{s}=\mu_0\int_s\mathbf{J}\cdot d\mathbf{a}+\mu_0\;\epsilon_0\;\frac{d}{dt}\int_s\mathbf{E}\cdot d\mathbf{a}$$

In summary,

- Gauss' law

$$\int_s\mathbf{E}\cdot d\mathbf{a}=\frac{1}{\epsilon_0}\int\rho dV$$

- Gauss' law for magnetism

$$\int_s\mathbf{B}\cdot d\mathbf{a}=0$$

- Faraday's law

$$\int_{loop}\mathbf{E}\cdot d\mathbf{s}=-\frac{d}{dt}\int_s\mathbf{B}\cdot d\mathbf{a}$$

- Ampère's law

$$\int_{loop}\mathbf{B}\cdot d\mathbf{s}=\mu_0\int_s\mathbf{J}\cdot d\mathbf{a}+\epsilon_0\;\mu_0\;\frac{d}{dt}\int_s\mathbf{E}\cdot d\mathbf{a}$$

In their integral form, Maxwell's equations can be used to make statements about a region of charge or current.

Differential Form

To make local statements and evaluate Maxwell's equations at individual points in space, one can recast Maxwell's equations in their differential form, which use the differential operators div and curl.

Differential form of Gauss' law. The divergence theorem holds that a surface integral over a closed surface can be written as a volume integral over the divergence inside the region. Thus

$$\frac{1}{\epsilon_0}\iiint\rho dV=\int_s\mathbf{E}\cdot d\mathbf{a}=\iiint\nabla\cdot\mathbf{E}dV$$

Since the statement is true for all closed surfaces, it must be the case that the integrands are equal, and thus

$$\nabla\cdot\mathbf{E}=\frac{\rho}{\epsilon_0}.$$

(The derivation of the differential form of Gauss' law for magnetism is identical.)

Differential form of Ampère's law. One can use Stokes' theorem to rewrite the line integral $\int \mathbf{B}.d\mathbf{s}$ in terms of the surface integral of the curl of $\mathbf{B}$:

$$\int_{loop} \mathbf{B}.d\mathbf{s} = \int_{surface} \nabla \times \mathbf{B} \cdot d\mathbf{a}$$

Ampère's law says that

$$\int_{loop} \mathbf{B}.d\mathbf{s} = \mu_o \int_s \mathbf{J} \cdot d\mathbf{a} + \mu_o \, \epsilon_o \frac{d}{dt} \int_s \mathbf{E} \cdot d\mathbf{a}.$$

Of course, the surface integral in both equations can be taken over any chosen closed surface, so the integrands must be equal:

$$\nabla \times \mathbf{B} = \mu_o \mathbf{J} + \mu_o \, \epsilon_o \frac{\partial \mathbf{E}}{\partial t}.$$

Differential form of Faraday's law. It follows from the integral form of Faraday's law that

$$\int_{loop} \mathbf{E} \cdot d\mathbf{s} = -\frac{d}{dt} \int_s \mathbf{B}.d\mathbf{a}$$

As was done with Ampère's law, one can invoke Stokes' theorem on the left side to equate the two integrands:

$$\int_s \nabla \times \mathbf{E} \cdot d\mathbf{a} = \frac{d}{dt} \int_s \mathbf{B}.d\mathbf{a}.$$

Again, one argue that since the relationship must hold true for any arbitrary surface S, it must be the case that the two integrands are equal, and therefore

$$\nabla \times \mathbf{E} = -\frac{d\mathbf{B}}{dt}.$$

In all, the result is as follows:

- Gauss' law

$$\nabla \cdot \mathbf{E} = -\frac{\rho}{\epsilon_o}$$

- Gauss' law for magnetism

$$\nabla \cdot \mathbf{B} = 0$$

- Faraday's law

$$\nabla \times \mathbf{E} = -\frac{d\mathbf{B}}{dt}$$

- Ampère's law

$$\nabla \times \mathbf{B} = \mu_o \mathbf{J} + \mu_o \in_o \frac{\partial \mathbf{E}}{\partial t}$$

Electromagnetic Waves

By assembling all four of Maxwell's equations together and providing the correction to Ampère's law, Maxwell was able to show that electromagnetic fields could propagate as traveling waves. In other words, Maxwell's equations could be combined to form a wave equation. Maxwell's insight stands as one of the great theoretical triumphs of physics. A simple sketch of this result is as follows.

For simplicity, suppose there is some region of space in which the electric field $E(x)$ is non-zero only along the z axis and the magnetic field $B(x)$ is non-zero only along the y axis, such that both are functions of x only. Then Faraday's law gives

$$\frac{\partial E}{\partial x} = -\frac{\partial B}{\partial t}.$$

Even though $J = 0$, with the additional term, Ampere's law now gives

$$\frac{\partial B}{\partial x} = -\frac{1}{c^2}\frac{\partial E}{\partial t}$$

Taking the partial derivative of the first equation with respect to x and the second with respect to t yields

$$\frac{\partial^2 E}{\partial x^2} = -\frac{\partial^2 B}{\partial x \partial t}$$

$$\frac{\partial^2 B}{\partial t \partial x} = -\frac{1}{c^2}\frac{\partial^2 E}{\partial t^2}$$

Therefore

$$\frac{\partial^2 E}{\partial x^2} = \frac{1}{c^2}\frac{\partial^2 E}{\partial t^2}$$

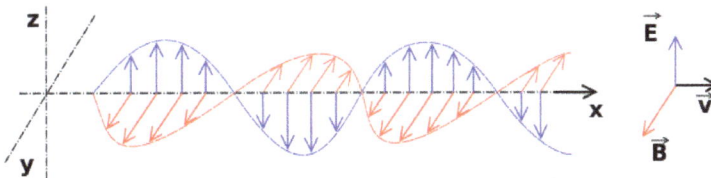

This equation has solutions for $E(x)$ (and corresponding solutions for $B(x)$ that represent traveling electromagnetic waves. In fact, the equation that has just been derived is in fact in the same form as the classical wave equation in one dimension. In other words, the laws of electricity and magnetism permit for the electric and magnetic fields to travel as waves, but only if Maxwell's correction is added to Ampère's law. Indeed, Maxwell was the first to provide a theoretical explanation of a classical electromagnetic wave and, in doing so, compute the speed of light.

Reference

- Basic-properties-electric-charge, electric-charges-and-fields, physics: toppr.com, Retrieved 14 April, 2019

- Electric-current, science: britannica.com, Retrieved 13 June, 2019

- Electric-current: circuitglobe.com, Retrieved 3 February, 2019

- What-is-electrical-current, current: electronics-notes.com, Retrieved 17 May, 2019

- Electric-current-and-circuit-diagrams, electricity, physics: toppr.com, Retrieved 27 January, 2019

- Electrical-force, physics: byjus.com, Retrieved 29 April, 2019

- Coulombs-law, electric-charges-and-fields, physics: toppr.com, Retrieved 1 March, 2019

- Electrostatics, science: siyavula.com, Retrieved 7 March, 2019

- Electric-field: circuitglobe.com, Retrieved 18 January, 2019

- Electric-potential: electrical4u.com, Retrieved 14 May, 2019

- Electric-flux-and-gausss-law, boundless-physics: lumenlearning.com, Retrieved 22 July, 2019

- Gauss-law: byjus.com, Retrieved 10 April, 2019

- What-is-magnetic-force, magnets-magnetic, magnetic-forces-and-magnetic-fields, physics, science: khanacademy.org, Retrieved 19 June, 2019

- Magnetic-field: techtarget.com, Retrieved 23 August, 2019

- Magnetic-force-and-magnetic-field, moving-charges-and-magnetism, physics: toppr.com, Retrieved 26 February, 2019

- Amperes_law_formula, physics, formulas: softschools.com, Retrieved 6 March, 2019

- Faradays-law, physics: byjus.com, Retrieved 28 July, 2019

- Lenz-law-of-electromagnetic-induction: electrical4u.com, Retrieved 8 February, 2019

- Maxwells-equations: brilliant.org, Retrieved 17 April, 2019

6

Optics: The Study of Light

The branch of physics that studies the properties and behaviour of light is known as optics. It studies its interactions with matter and the construction of instrument that detects it. The chapter closely examines the key concepts of optics to provide an extensive understanding of the subject.

Light

Light is part of the electromagnetic spectrum, which ranges from radio waves to gamma rays. Electromagnetic radiation waves, as their names suggest are fluctuations of electric and magnetic fields, which can transport energy from one location to another. Visible light is not inherently different from the other parts of the electromagnetic spectrum with the exception that the human eye can detect visible waves. Electromagnetic radiation can also be described in terms of a stream of photons which are massless particles each travelling with wavelike properties at the speed of light. A photon is the smallest quantity (quantum) of energy which can be transported and it was the realization that light travelled in discrete quanta that was the origins of Quantum Theory.

It is no accident that humans can 'see' light. The detection of light is a very powerful tool for probing the universe around us. As light interacts with matter it can be become altered and by studying light that has originated or interacted with matter, many of the properties of that matter can be determined. It is through the study of light that for example we can understand the composition of the stars light years away or watch the processes that occur in the living cell as they happen.

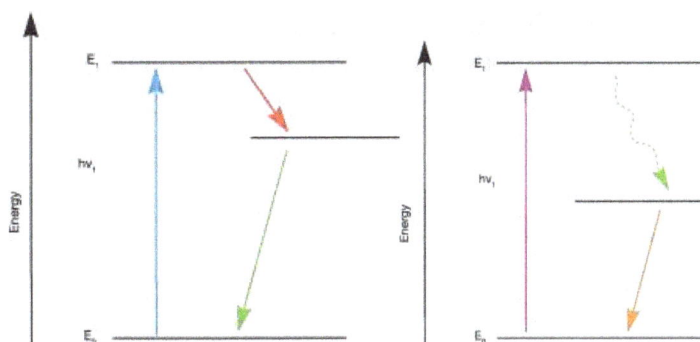

Matter is composed of atoms, ions or molecules and it is light's interaction with matter which gives rise to the various phenomena which can help us understand the nature of matter. The atoms, ions or molecules have defined energy levels usually associated with energy levels that electrons in the matter can hold. Light can be generated by the matter or a photon of light can interact with the energy levels in a number of ways.

We can represent the energy levels in a diagram known as a Jablonski diagram. An example of one is shown in the diagram above. An atom or molecule in the lowest energy state possible known as the ground state can absorb a photon which will allow the atom or molecule to be raised to a higher energy level state or become excited. Hence the matter can absorb light of characteristic wavelengths such as the blue light in the example on the right or the violet light in the example on the left. The atom or molecule won't stay in an excited state so it relaxes back to the ground state by several ways. In the example on the right, the atom or molecule emits two photons both of lower energy than the absorbed photon. The photons emitted will be a characteristic energy appropriate for a particular atom or compound and so by studying the light emission the matter under investigation can be determined. In the example on the left the excited atom or molecule initially loses energy by not emitting a photon and instead relaxes to the lower energy state by internal processes which typically heat up the matter. The intermediate energy level then relaxes to the ground state by the emission of a photon of orange light.

Optics

Optics is the science concerned with the genesis and propagation of light, the changes that it undergoes and produces, and other phenomena closely associated with it. There are two major branches of optics, physical and geometrical. Physical optics deals primarily with the nature and properties of light itself. Geometrical optics has to do with the principles that govern the image-forming properties of lenses, mirrors, and other devices that make use of light. It also includes optical data processing, which involves the manipulation of the information content of an image formed by coherent optical systems.

Originally, the term optics was used only in relation to the eye and vision. Later, as lenses and other devices for aiding vision began to be developed, these were naturally called optical instruments, and the meaning of the term optics eventually became broadened to cover any application of light, even though the ultimate receiver is not the eye but a physical detector, such as a photographic plate or a television camera. In the 20th century optical methods came to be applied extensively to regions of the electromagnetic radiation spectrum not visible to the eye, such as X-rays, ultraviolet, infrared, and microwave radio waves, and to this extent these regions are now often included in the general field of optics.

Geometrical Optics

Geometrical optics, or ray optics, is frequently used to study how images form in optical systems. It uses the concept of rays, which have direction and position but no phase information, to model the way Electromagnetic Radiation (Electromagnetic Radiation) travels through an optical system. Geometrical Optics is an approximation of how EM Rad behaves in an optical system – it is a very good model to use when the smallest dimension of the optical system is much larger than the wavelength of the incident EM Rad. But if the dimensions of the optical components are about the size of a wavelength or smaller, a different model (either the Electromagnetic Wave Optics Model or the Quantum Optics Model) must be used.

In geometrical optics, an object is viewed as a collection of many pin-point sources of EM Rad. Each source produces a bundle of rays that is traced through an optical system to determine what image will be formed. This technique is useful for designing lenses, developing systems of lenses for use in microscopes, telescopes, cameras and other devices, and for studying the aberrations of an optical system.

Reflection and Refraction

Reflection is the phenomenon of bouncing back of light into the same medium on striking the surface of any object.

Laws of Reflection

- First law: The incident ray, the normal to the surface at the point of incidence and the reflected ray, all lie in the same plane.

- Second law: The angle of reflection (r) is always equal to the angle of incidence (i).

   $\angle i = \angle r$

The image formed by a plane mirror is always

- virtual and erect
- of the same size as the object
- as far behind the mirror as the object is in front of it
- laterally inverted

Spherical mirrors are of two types:

- Convex mirrors or diverging mirrors in which the reflecting surface is curved outwards.
- Concave mirrors or converging mirrors in which the reflecting surface is curved inwards.

Some terms related to spherical mirrors:

- The centre of curvature (C) of a spherical mirror is the centre of the hollow sphere of glass, of which the spherical mirror is a part.

- The radius of curvature (R) of a spherical mirror is the radius of the hollow sphere of glass, of which the spherical mirror is a part.

- The pole (P) of a spherical mirror is the centre of the mirror.

- The principal axis of a spherical mirror is a straight line passing through the centre of curvature C and pole P of the spherical mirror.

- The principal focus (F) of a concave mirror is a point on the principal axis at which the rays of light incident on the mirror, in a direction parallel to the principal axis, actually meet after reflection from the mirror.

- The principal focus (F) of a convex mirror is a point on the principal axis from which the rays of light incident on the mirror, in a direction parallel to the principal axis, appear to diverge after reflection from the mirror.

- The focal length (f) of a mirror is the distance between its pole (P) and principal focus (F).

- For spherical mirrors of small aperture, R = 2f.

Sign Conventions for Spherical Mirrors

According to New Cartesian Sign Conventions,

- All distances are measured from the pole of the mirror.

- The distances measured in the direction of incidence of light are taken as positive and vice versa.

- The heights above the principal axis are taken as positive and vice versa.

Rules for Tracing Images Formed by Spherical Mirrors

Rule 1: A ray which is parallel to the principal axis after reflection passes through the principal focus in case of a concave mirror or appears to diverge from the principal focus in case of a convex mirror.

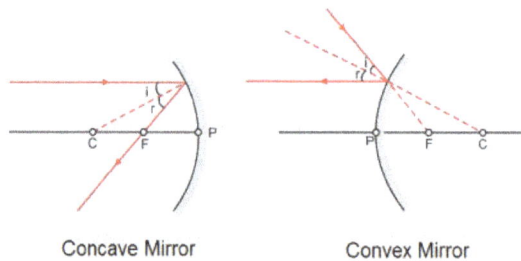

Concave Mirror   Convex Mirror

Rule 2: A ray passing through the principal focus of a concave mirror or a ray which is directed towards the principal focus of a convex mirror emerges parallel to the principal axis after reflection

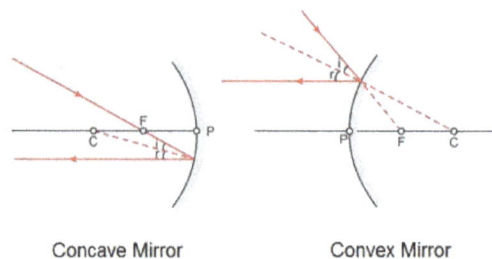

Concave Mirror   Convex Mirror

Rule 3: A ray passing through the centre of curvature of a concave mirror or directed towards the centre of curvature of a convex mirror is reflected back along the same path.

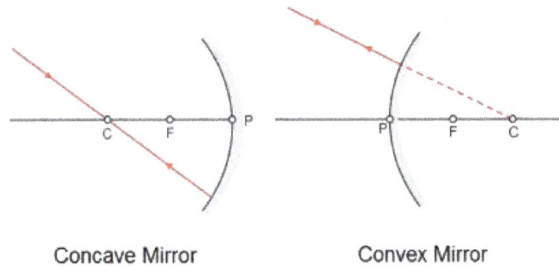

Concave Mirror   Convex Mirror

Rule 4: A ray incident obliquely towards the pole of a concave mirror or a convex mirror is reflected obliquely as per the laws of reflection.

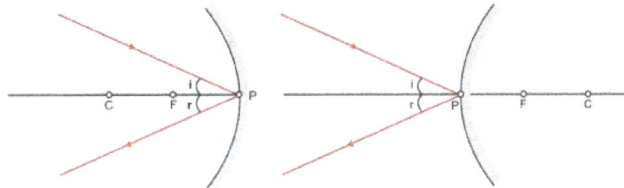

Image Formation by a Concave Mirror

- Ray Diagrams

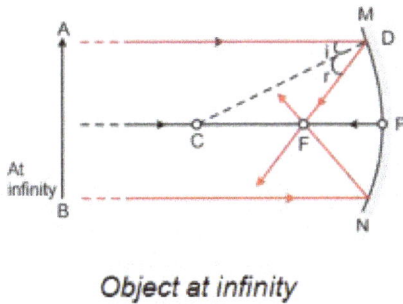

Object at infinity

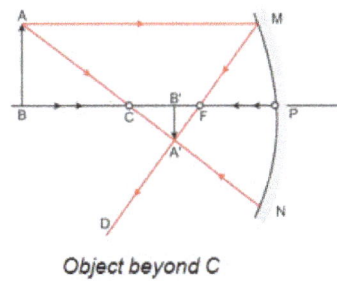

Object beyond C

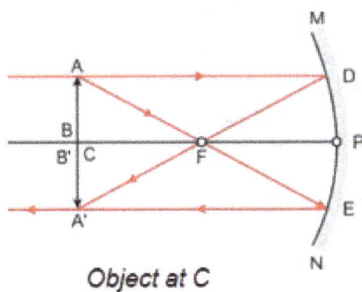

Object at C

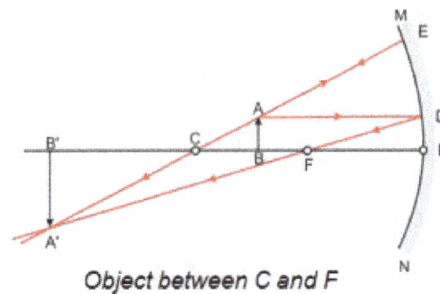

Object between C and F

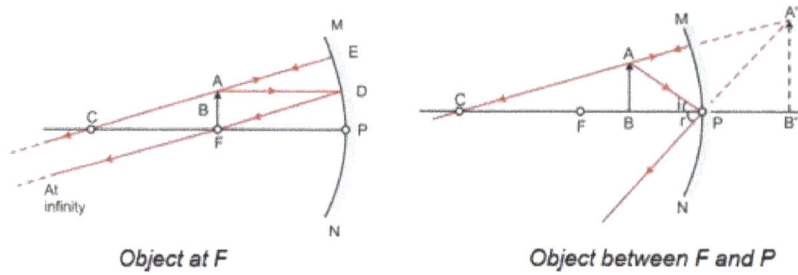

Object at F   *Object between F and P*

• Characteristics of images formed

Position of object	Position of image	Size of image	Nature of image
At infinity	At focus F	Highly diminished	Real and inverted
Beyond C	Between F and C	Diminished	Real and inverted
At C	At C	Equal to size of object	Real and inverted
Between C and F	Beyond C	Enlarged	Real and inverted
At F	At infinity	Highly enlarged	Real and inverted
Between F and P	Behind the mirror	Enlarged	Virtual and erect

Image Formation by a Convex Mirror

• Ray Diagrams

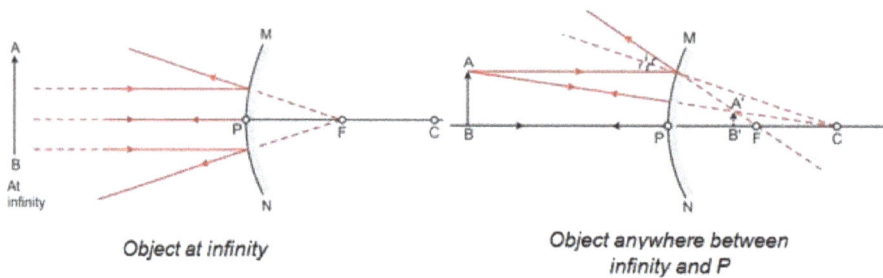

Object at infinity   *Object anywhere between infinity and P*

• Characteristics of images formed

Position of object	Position of image	Size of image	Nature of image
At infinity	At focus F behind the mirror	Highly diminished, point sized	Virtual and erect
Anywhere between infinity and the pole of the mirror	Between P and F behind the mirror	Diminished	Virtual and erect

Mirror Formula

The object distance (u), image distance (v) and focal length (f) of a spherical mirror are related as,

$$\frac{1}{u} + \frac{1}{V} = \frac{1}{f}$$

Linear Magnification (m) produced by a spherical mirror is,

$$m = \frac{\text{size of image } (h_2)}{\text{size of object } (h_1)} = -\frac{\text{image distance } (v)}{\text{object distance } (u)}$$

m is negative for real images and positive for virtual images.

Refraction of Light

The phenomenon of change in the path of a beam of light as it passes from one medium to another is called refraction of light.

The cause of refraction is the change in the speed of light as it goes from one medium to another.

Laws of Refraction

- First Law: The incident ray, the refracted ray and the normal to the interface of two media at the point of incidence, all lie in the same plane.

- Second Law: The ratio of the sine of the angle of incidence to the sine of the angle of refraction is constant for a given pair of media.

$$\frac{\sin i}{\sin r} = \text{constant} = {}^1n_2$$

This law is also known as Snell's law.

The constant, written as $1n$ is called the refractive index of the second medium (in which the

refracted ray lies) with respect to the first medium (in which the incident ray lies).

Absolute refractive index (n) of a medium is given as,

$$n = \frac{\text{speed of light in vacuum}}{\text{speed of light in the medium}} = \frac{c}{v}$$

When a beam of light passes from medium 1 to medium 2, the refractive index of medium 2 with respect to medium 1 is called the relative refractive index, represented by ${}^1n_2$, where

$$^1n_2 = \frac{n_2}{n_1} = \frac{c/v_2}{c/v_1} = \frac{v_1}{v_2}$$

Similarly, the refractive index of medium 1 with respect to medium 2 is,

$$^2n_1 = \frac{n_1}{n_2} = \frac{c/v_1}{c/v_2} = \frac{v_2}{v_1}$$

$$\Rightarrow \qquad {}^1n_2 \times {}^2n_1 = 1$$

$$\text{or.} \qquad {}^1n_2 = \frac{1}{{}^2n_1}$$

While going from a rarer to a denser medium, the ray of light bends towards the normal.

While going from a denser to a rarer medium, the ray of light bends away from the normal.

Conditions for No Refraction

- When light is incident normally on a boundary.

- When the refractive indices of the two media are equal.

- In the case of a rectangular glass slab, a ray of light suffers two refractions, one at the air–glass interface and the other at the glass–air interface. The emergent ray is parallel to the direction of the incident ray.

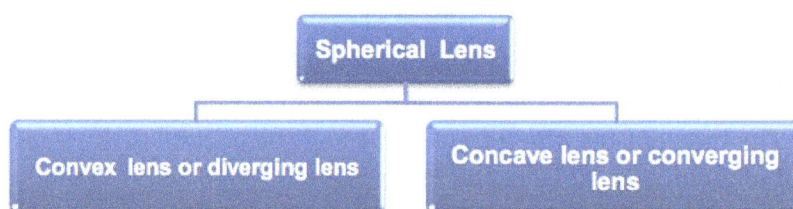

- Convex lens or converging lens which is thick at the centre and thin at the edges.

- Concave lens or diverging lens which is thin at the centre and thick at the edges.

Some terms related to spherical lenses

- The central point of the lens is known as its optical centre (O).

- Each of the two spherical surfaces of a lens forms a part of a sphere. The centres of these spheres are called centres of curvature of the lens. These are represented as C_1 and C_2.

- The principal axis of a lens is a straight line passing through its two centres of curvature.

- The principal focus of a convex lens is a point on its principal axis to which light rays parallel to the principal axis converge after passing through the lens.

- The principal focus of a concave lens is a point on its principal axis from which light rays.

- Originally parallel to the principal axis appear to diverge after passing through the lens.

- The focal length (f) of a lens is the distance of the principal focus from the optical centre.

Sign Conventions for Spherical Lenses

According to new cartesian sign conventions,

- All distances are measured from the optical centre of the lens.

- The distances measured in the direction of incidence of light are taken as positive and vice versa.

- The heights above the principal axis are taken as positive and vice versa.

Rules for Tracing Images formed by Spherical Lens

Rule 1: A ray which is parallel to the principal axis, after refraction passes through the principal focus on the other side of the lens in case of a convex lens or appears to diverge from the principal focus on the same side of the lens in case of a concave lens.

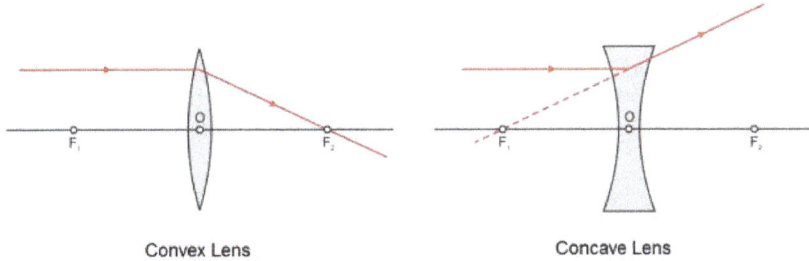

Convex Lens   Concave Lens

Rule 2: A ray passing through the principal focus of a convex lens or appearing to meet at the principal focus of a concave lens after refraction emerges parallel to the principal axis.

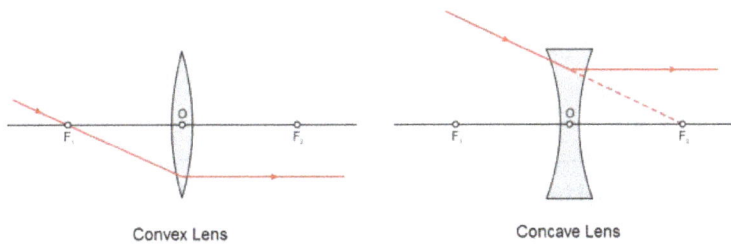

Convex Lens   Concave Lens

Rule 3: A ray passing through the optical centre of a convex lens or a concave lens emerges without any deviation

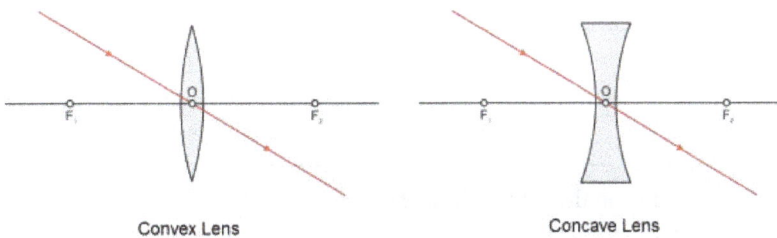

Convex Lens   Concave Lens

Image Formation by a Convex Lens

- Ray Diagrams

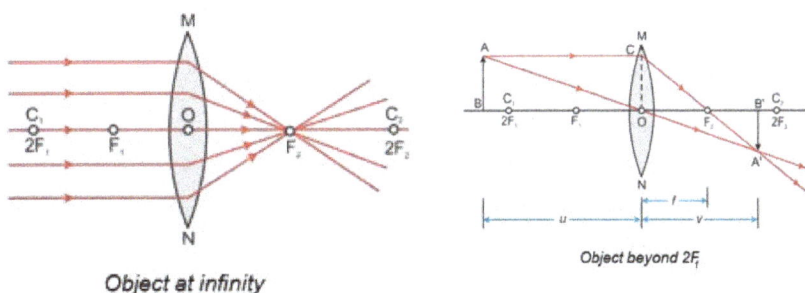

Object at infinity

Object beyond 2F₁

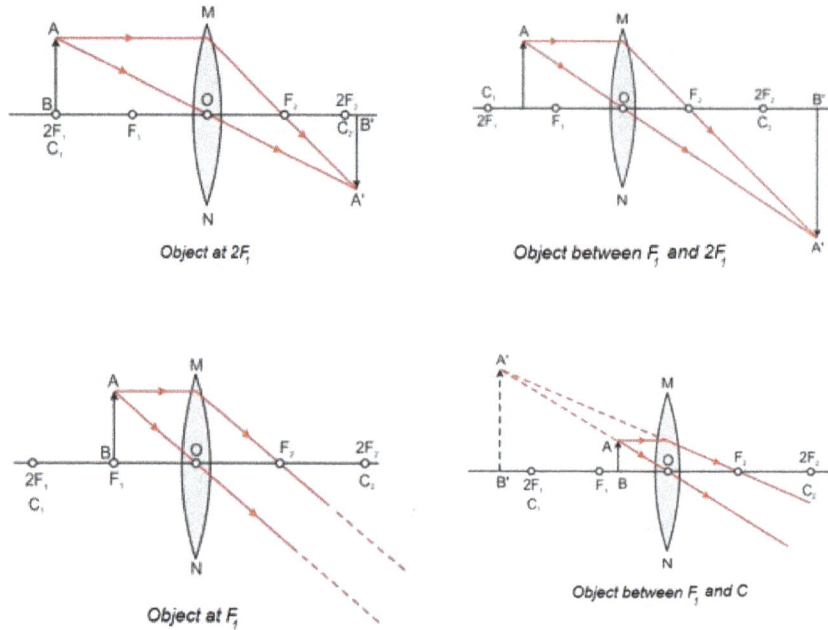

Object at 2F₁

Object between F₁ and 2F₁

Object at F₁

Object between F₁ and C

- Characteristics of images formed

Position of object	Position of image	Size of image	Nature of image
At infinity	At focus F_2	Highly diminished	Real and inverted
Beyond $2F_1$	Between F_2 and $2F_2$	Diminished	Real and inverted
At $2F_1$	At $2F_2$	Equal to size of object	Real and inverted
Between F_1 and $2F_1$	Beyond $2F_2$	Enlarged	Real and inverted
At focus F_1	At infinity	Highly enlarged	Real and inverted
Between F_1 and O	Beyond F_1 on the same side as the object	Enlarged	Virtual and erect

Image Formation by a Concave Lens

- Ray Diagrams

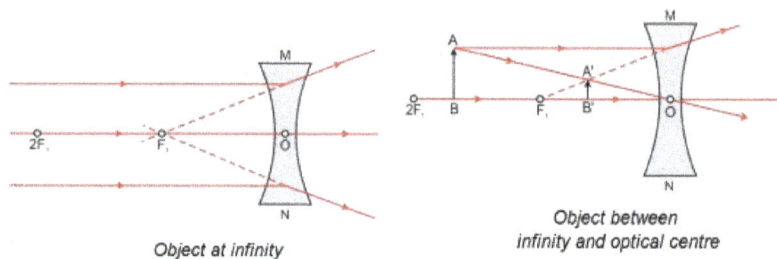

Object at infinity

Object between
infinity and optical centre

- Characteristics of images formed

Position of object	Position of image	Size of image	Nature of image
At infinity	At focus F_1	Highly diminished	Virtual and erect
Between infinity and O	Between focus F_1 and O	Diminished	Virtual and erect

Lens Formula

Object distance (u), image distance (v) and focal length (f) of a spherical lens are related as $\dfrac{1}{v} - \dfrac{1}{u} = \dfrac{1}{f}$

Linear Magnification (m) produced by a spherical lens is,

$$m = \frac{\text{size of image } (h_2)}{\text{size of object } (h_1)} = \frac{\text{image distance } (v)}{\text{object distance } (u)}$$

m is negative for real images and positive for virtual images.

Power of a Lens

- Power of a lens is the reciprocal of the focal length of the lens. Its S.I. unit is dioptre (D).

$$P \text{ (dioptre)} = \frac{1}{f \text{(metre)}}$$

- Power of a convex lens is positive and that of a concave lens is negative.

- When several thin lenses are placed in contact with one another, the power of the combination of lenses is equal to the algebraic sum of the powers of the individual lenses.

$$P = P_1 + P_2 + P_3 + P_4 + \dots$$

Wave Optics

Wave optics deals with the connection of waves and rays of light. It is used when the wave characteristics of light are taken in account. Wave optics deals with the study of various phenomenal behaviors of light like reflection, refraction, interference, diffraction, polarization etc. It is otherwise known as physical optics.

For example, consider a compact disc. The colors which is reflected by the disc varies with different angle. This may be the evidence for wave character of light. It happens due to the interference of light waves which can be described by the concept of wave optics. It's the same concept we see when we look at soap bubbles and a thin film of oil which is floating on surface of water. This reflection phenomena can be explained with the wave theory of light. Another example to explain this phenomenon is peacock and butterfly.

Evidence for the wave character of light

Wave optics explains why the sky is blue. The white light which is emitted by the sun is actually made up of all colors which are present in the beautiful rainbow. As per wave theory of light, light travels in wave forms too. Some light moves in long waves whereas, some other moves in the form of short waves. Here blue light waves are shorter than red light waves. All the light colors including blue light reaches the earth's atmosphere and is scattered in all directions. As blue light is shorter and smaller wave, it is scattered more than any other colors in to the earth's atmosphere. This is the reason for the blue colour of sky.

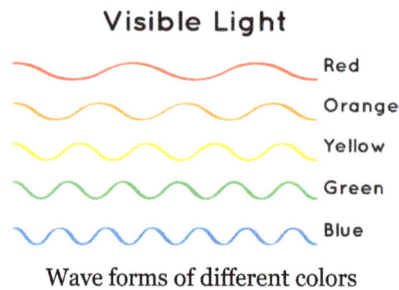

Wave forms of different colors

Wavefront

A wave front is a surface or line in the path of wave motion on which the disturbances at every point have the same phase.

Depending upon the source of light, wave fronts can be of three types: spherical wave front, cylindrical wave front and plane wave front.

Spherical Wavefront

If a point source in an isotropic medium is sending out waves in three dimensions, the wave fronts are spheres centered on the source as shown in figure. Such wave front is called a spherical wave front.

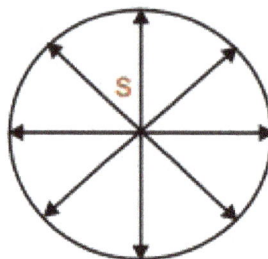

Cylindrical Wavefront

When the source of light is linear, all the points equidistant from the linear source lie on the surface of a cylinder as shown in figure. Such a wave front is called a cylindrical wave front.

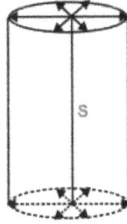

Plane Wavefront

At a large distance from a source of any kind, the wave front will appear plane as shown in figure. Such a wave front is called plane wave front.

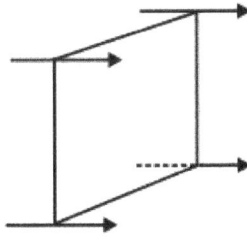

Huygen's Principle

In optics, Huygens' principle is a statement that all points of a wave front of light in a vacuum or transparent medium may be regarded as new sources of wavelets that expand in every direction at a rate depending on their velocities. Proposed by the Dutch mathematician, physicist, and astronomer, Christiaan Huygens, in 1690, it is a powerful method for studying various optical phenomena.

A surface tangent to the wavelets constitutes the new wave front and is called the envelope of the wavelets. If a medium is homogeneous and has the same properties throughout (i.e., is isotropic), permitting light to travel with the same speed regardless of its direction of propagation, the three-dimensional envelope of a point source will be spherical; otherwise, as is the case with many crystals, the envelope will be ellipsoidal in shape. An extended light source will consist of an infinite number of point sources and may be thought of as generating a plane wave front.

So if we consider a point source, it will emit its wavefront and nature of the wavefront will be spherical one.

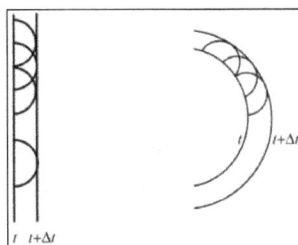

As per the Huygen's principle, all the points on the wave front are going to become a secondary source. So the wave fronts will in the forward direction. All the secondary sources emit wavelets. Tangent drawn to all the wavelets is the new position of the waveform.

This means that, suppose you are standing on the mountain and you throw a stone in the water from a height. What do you observe? You see that the stone strikes the surface of the water and waves are seen surrounding that point. Every point on the surface of water starts oscillating.

The waves spread in all the direction. Earlier the water was at rest. But the moment we throw the stone in the water, within a few fractions of seconds the disturbance spreads in all directions. There are ripples formed in the water. The ripples form the concentric circle around the disturbance and spread out.

These ripples are nothing but the wave front. The wave fronts gradually spread in all the directions. So at every point, we have a wave coming out. The primary wave front is formed and again from the primary wave front, a secondary waveform is formed and so on. The disturbance does not last for a long time. It fades gradually because more and more waveforms are formed.

Interference

One of the fundamental properties of light is its ability to interfere with itself. Most people observe optical interferences on a daily basis, but don't quite know how this phenomenon actually occurs. Some examples that people can relate to is a film of oil on water or a soap bubble that reflects a variety of beautiful colors when natural or artificial light is shone upon it. This dynamic interplay of colors derives from the simultaneous reflection of light from both the inside and outside surfaces of the bubble. The two surfaces are very close together (only a few microns thick) and light reflected from the inner surface interferes both constructively and destructively.

Principle of Superposition

The superposition principle is one of those ideas that sounds much more complicated than it really is. Physics pans out like that sometimes. The superposition principle states that for linear systems, the net response caused by two or more inputs is the sum of the inputs that each would have caused on its own. Probably, so let's simplify this a bit. In most cases, when people talk about the superposition principle, they're talking about waves or sinusoidal vibrations in space and time. Examples of waves include light, sound, water ripples, and earthquake waves. All of these things work in the same basic way: If you take two waves and put them on top of each other (or superimpose them) they add together. This is what superposition is.

When two waves are on top of each other, they combine to produce a total wave, which we call a resultant wave. We call it that because it's the result you get when the waves are added up. Waves contain peaks and troughs that come in a pattern, one after another. When you superimpose the peaks of two waves, they add together to form an even bigger peak. When you superimpose the troughs of two waves, they add together to form an even bigger trough. This is called constructive interference. On the other hand, when you superimpose the peak of one wave with the trough of another, they add together and flatten out to nothing—a flat line. This is called destructive interference. It's similar to how -6 + 6 = 0. The peak and trough cancel each other out.

Constructive and Destructive Interference

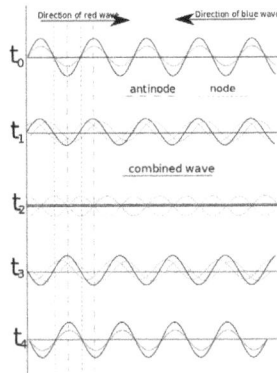

Most of the time, when we think about waves, we tend to imagine a single wave traveling through a medium. When we think about water waves, for example, we imagine a single wave traveling through the vastness of the ocean all by itself, but obviously, that's unrealistic. There are countless waves traveling in all directions. Some ocean waves are bigger and some are smaller. Some waves are caused by the wind, while others are caused by cruise ships, breaching whales and thousands of other things! Inevitably, some waves are going to cross over or meet with each other. When they do, the reaction between the waves is known as interference. This is the meeting of two or more waves traveling in the same medium. Waves meeting in the same medium actually disrupt each other's displacement. They interfere with each other so that the resulting wave is a completely new and different wave from either of the original two.

Imagine that they're traveling toward each other in the same medium. One is traveling left and the other is traveling right. They both have the same amplitude of 1 meter. When the two waves meet, there comes a moment when the crests of both waves end up in the same spot. Their crests overlap, so their amplitudes add together. Instead of the crest being 1-meter-tall, it's now 2 meters tall! When the crests or troughs of two interfering waves meet, their amplitudes add together. This is constructive interference. So, what happens when the crest of one wave meets the trough of another wave? Well, the opposite happens, and it's called destructive interference. When the crest and trough of two interfering waves meet, one amplitude subtracts from the other.

Let's take the same two waves that we considered above. They're still traveling toward each other and they are still 1 meter in amplitude each. However, this time, it just so happens that the crest of one wave lines up with the trough of the other wave. The crest of the first wave will cancel out the trough of the second wave. The medium experiences zero displacement and the net result is a completely flat surface.

Condition of a Steady Interference Pattern

   i. $A_1 = A_2$. The amplitude of two waves must be equal.

   ii. $\lambda_1 = \lambda_2$. The two waves interfering must have same color i.e they must be of the same wavelength.

  iii. Sources must be narrow.

iv. The distance between source should be less.

v. Source and screen should be at large distance.

vi. We should get coherent sources.

Path difference for constructive and destructive interference:

Suppose there are two coherent sources S_1 and S_2. There is also a point source P. The point source P is located at the same distance from the sources S_1 and S_2. When both the sources are in the same phase, the constructive path difference will be $0, \lambda, 2\lambda$.......The destructive path difference will be $\lambda/2, 3\lambda/2, 5\lambda/2$......

Young's Double Slit Experiment

Young's double-slit experiment uses two coherent sources of light placed at a small distance apart, usually, only a few orders of magnitude greater than the wavelength of light is used. Young's double-slit experiment helped in understanding the wave theory of light which is explained with the help of a diagram. A screen or photodetector is placed at a large distance 'D' away from the slits as shown.

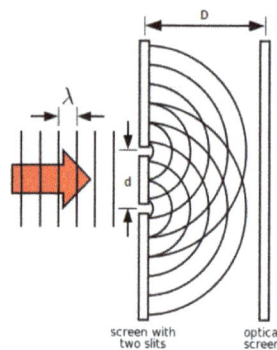

Derivation of Young's Double Slit Experiment

The original Young's double-slit experiment used diffracted light from a single source passed into two more slits to be used as coherent sources. Lasers are commonly used as coherent sources in the modern-day experiments.

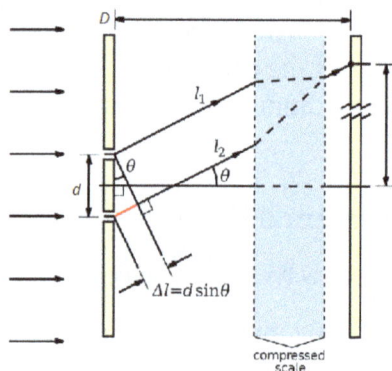

Young's Double Slit Experiment

Each source can be considered as a source of coherent light waves. At any point on the screen at a distance 'y' from the centre, the waves travel distances l_1 and l_2 to create a path difference of Δl at the point in question. The point approximately subtends an angle of θ at the sources (since the distance D is large there is only a very small difference of the angles subtended at sources).

Derivation of Young's Double Slit Experiment

Consider a monochromatic light source's' kept at a considerable distance from two slits s_1 and s_2. S is equidistant from s1 and s_2. s_1 and s_2 behave as two coherent sources, as both bring derived from S.

The light passes through these slits and falls on a screen which is at a distance 'D' from the position of slits s_1 and s_2. 'd' be the separation between two slits.

If s1 is open and s2 is closed, the screen opposite to s1 is closed, only the screen opposite to s2 is illuminated. The interference patterns appear only when both slits s_1 and s_2 are open.

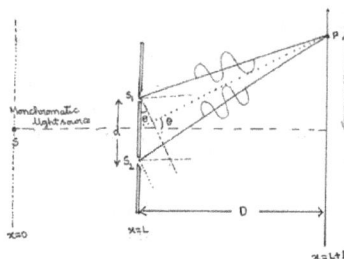

When the slit separation (d) and the screen distance (D) are kept unchanged, to reach P the light waves from s_1 and s_2 must travel different distances. It implies that there is a path difference in Young's double slit experiment between the two light waves from s_1 and s_2.

Approximations in Young's Double Slit Experiment

- Approximation 1:D > > d: Since D > > d, the two light rays are assumed to be parallel, then the path difference,

- Approximation 2: d/λ >> 1: Often, d is a fraction of a millimetre and λ is a fraction of a micrometre for visible light.

Under these conditions θ is small, thus we can use the approximation $\sin \theta$ approx $\tan \theta = y/D$.

$\therefore$ path difference, $\Delta z = y/D$

This is the path difference between two waves meeting at a point on the screen. Due to this path difference in Young's double slit experiment, some points on the screen are bright and some points are dark.

Now, we will discuss the position of these light, dark fringes and fringe width.

Position of Fringes in Young's Double Slit Experiment

Position of Bright Fringes

For maximum intensity at P

$$z = n \quad (n = 0, \pm1, \pm2, \ldots)$$

Or $d \sin \theta = n\lambda \quad (n = 0, \pm1, \pm2, \ldots)$

The bright fringe for n = 0 is known as the central fringe. Higher order fringes are situated symmetrically about the central fringe. The position of nth bright fringe is given by,

$$y \, (\text{bright}) = (n\lambda \backslash d)D \ (n = 0, \pm1, \pm2, \ldots)$$

Position of Dark Fringes

For minimum intensity at P,

$$\Delta z = (2n-1)\frac{n\lambda}{2} (n = \pm1, \pm2, \ldots) d \sin \theta = (2n-1)\frac{n\lambda}{2}$$

The first minima are adjacent to the central maximum on either side. We will obtain the position of dark fringe as,

$$y_{dark} = \frac{(2n-1)\lambda D}{2d} (n = \pm1, \pm2, \ldots)$$

Fringe Width

Distance between two adjacent bright (or dark) fringes is called the fringe width.

$$\beta z \frac{n\lambda D}{d} - \frac{(n-1)\lambda D}{d} = \frac{\lambda D}{d}$$
$$\Rightarrow \beta \propto \lambda$$

If the apparatus of Young's double slit experiment is immersed in a liquid of refractive index (u), then wavelength of light and hence fringe width decreases 'u' times.

$$\beta^1 = \frac{\beta}{\mu}$$

If a white light is used in place of monochromatic light, then coloured fringes are obtained on the screen with red fringes larger in size that of violet.

Angular Width of Fringes

Let the angular position of nth bright fringe is θ_n and because of its small value $\tan\theta n \approx \theta n$,

$$\tan \theta_n \approx \theta_n \tan\theta_n = \frac{yn}{D} \approx \theta_n = \frac{yn}{D}$$

$$\Rightarrow \theta_n = \frac{n\gamma D / d}{D} = \frac{n\lambda}{d}$$

Similarly, the angular position of $(n+1)^{th}$ bright fringe is θ_{n+1}, then

$$\theta_{n+1} = \frac{\gamma_{n+1}}{D} = \frac{(n+1)\lambda D / d}{D} = \frac{(n+1)\lambda}{d}$$

∴ Angular width of a fringe in Youngs double slit experiment is given by,

$$\theta = \theta_{n+1} - \theta_n = \frac{(n+1)\lambda}{d} - \frac{n\lambda}{d} = \frac{\lambda}{d} \; \theta = \frac{\lambda}{d}$$

We know that $\beta = \frac{\lambda D}{d}$

$$\Rightarrow \theta = \frac{\lambda}{d} = \frac{\beta}{D}$$

Angular width is independent of 'n' i.e angular width of all fringes are same.

Maximum Order of Interference Fringes

The position of nth order maxima on the screen is $\gamma = \frac{n\lambda D}{d}; n = 0, \pm 1, \pm 2, ..$

But 'n' values cannot take infinitely large values as it would violate 2nd approximation.

i.e θ is small (or) y << D

$$\Rightarrow \frac{\gamma}{D} = \frac{n\lambda}{d} << 1$$

Hence, the above formula for interference maxima is applicable when $n << \frac{d}{\lambda}$

When 'n' value becomes comparable to $\frac{d}{\lambda}$, path difference can no longer be given by $\frac{d\gamma}{D}$, Hence for maxima, path difference = nλ

$$\Rightarrow d_{\sin}\theta = n\lambda$$

$$\Rightarrow n = \frac{d_{\sin}\theta}{\lambda} \; n_{max} = \left(\frac{d}{\lambda}\right)$$

The above represents box function or greatest integer function.

Similarly, the highest order of interference minima

$$n_{min} = \left(\frac{d}{\lambda} + \frac{1}{2}\right)$$

Shape of Interference Fringes in YDSE

From the given YDSE diagram, the path difference from the two slits is given by,

$$s_2 p - s_1 p \approx d\sin\theta \text{ (constant)}$$

The above equation represents a hyperbola with its two foci as s_1 and s_2.

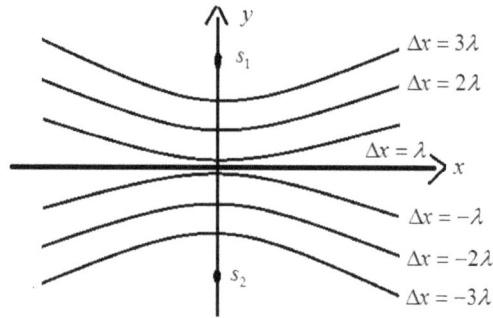

The interference pattern we get on the screen is a section of a hyperbola when we revolve hyperbola about the axis s_1s^2.

If the fringe will represent 1st minima, the fringe will represent 1st maxima, it represents central maxima.

If the screen is yz plane, fringes are hyperbolic with a straight central section.

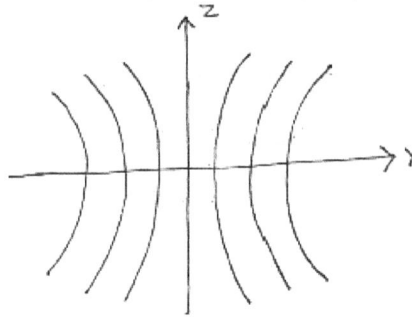

If the screen is xy plane, the fringes are hyperbolic with a straight central section.

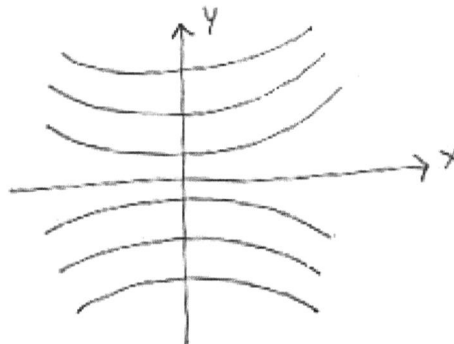

Intensity of Fringes in Young's Double Slit Experiment

For two coherent sources s1 and s2, the resultant intensity at point p is given by,

$$I = I1 + I2 + 2 \sqrt{(I1 . I2)} \cos \varphi$$

Putting $I1 = I2 = I0 \, (\text{Since, } d<<<D)$

$I = I0 + I0 + 2 \; \sqrt{(I0.I0)}\cos \varphi$

$I = 2I0 + 2 \; (I0) \cos \varphi$

$I = 2I0 \; (H \cos s \; \varphi)$

$I = 4I_0\cos^2\left(\dfrac{\phi}{2}\right).$

For Maximum Intensity

$\cos\dfrac{\phi}{2} = \pm 1 \dfrac{\phi}{2} = n\pi, n = 0,\pm 1,\pm 2,\ldots\ldots$

or $\varphi = 2n\pi$

phase difference $\varphi = 2n\pi$

then, path difference $\Delta x = \dfrac{\lambda}{2\pi}(2n\pi) = n\lambda$

The intensity of bright points are maximum and given by,

Imax = 4I0

For minimum intensity:

$\cos\dfrac{\phi}{2} = 0 \dfrac{\phi}{2} = (n-\dfrac{1}{2}) \, \pi \; where \; (n = \pm 1,\pm 2,\pm 3,\ldots..)$

$\varphi = (2n - 1) \, \pi$

Phase difference $\varphi = (2n - 1)\pi$

$\dfrac{2\pi}{\lambda}\Delta x = (2n-1)\pi \; \Delta x = (2n-1)\dfrac{\lambda}{2}$

Thus, intensity of minima is given by

Imin = 0

If $I_1 \neq I_2, I_{min} \neq 0.$

Special Cases

Rays Not Parallel to Principal Axis

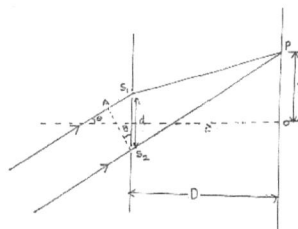

From the above diagram,

Path difference $\Delta x = (AS_1 + S_1P) - S_2P \,\, \Delta x = AS_1 - (S_2P - S_1P)$

$$\Delta x = d sin\theta - \frac{4}{dD}$$

For maxima $\Delta x = n\lambda$

For minima $\Delta x = (2n-1)\dfrac{\lambda}{2}$

Using this we can calculate different positions of maxima and minima.

Source Placed Beyond the Central Line

If the source is placed a little above or below this centre line, the wave interaction with S1 and S2 has a path difference at a point P on the screen,

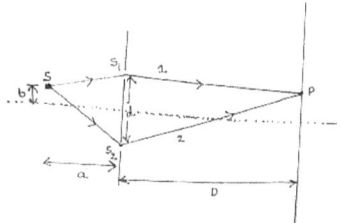

Δ x= (distance of ray 2) – (distance of ray 1)

$$= (SS_2 + S_2P) - (SS_1 + S_1P)$$
$$= (SS_2 + SS_1) + (S_2P - S_1P)$$

$$= \text{bd/a} + \text{yd/D} \rightarrow (*)$$

We know $\Delta x = n\lambda$ for maximum

$\Delta x = (2n-1) \,\, \lambda/2$ for minimum

By knowing the value of Δx from (*) we can calculate different positions of maxima and minima.

Displacement of Fringes in YDSE

When a thin transparent plate of thickness 't' is introduced in front of one of the slits in Young's double slit experiment, the fringe pattern shifts toward the side where the plate is present.

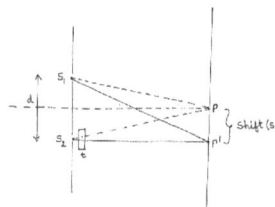

The dotted lines denote the path of the light before introducing the transparent plate. The solid lines denote the path of the light after introducing a transparent plate.

Path difference before introducing the plate $\Delta x = S_1 P - S_2 P$

Path difference after introducing the plate $\Delta x_{new} = S_1 P^1 - S_2 P^1$

The path length $S_2 P^1 = (S_2 P^1 - t)_{air} + t_{plat} = (S_2 P^1 - t)_{air} + (\mu t)_{plat}$

where 'μt' is optical path,

$$= S_2 P^1_{\;air} + (\mu - 1)t$$

Then We get

$$(\Delta x)_{new} = S_1 P^1_{\;air} - (S_2 \; P^1 air + (\mu - 1)t)$$

$$= (S_1 P^1 - S_2 \; P^1)_{air} + (\mu - 1)t (\Delta x)_{new} = d\, Sin\,\theta - (\mu - 1)t$$

$$(\Delta x)_{new} = \frac{\gamma d}{D} - (\mu - 1)t$$

Then,

$$y = \frac{\Delta x D}{d} + \frac{D}{d}\left[((\mu - 1)t)\right]$$

$$\downarrow \qquad\qquad \downarrow$$

$$(1) \qquad\qquad (2)$$

Term (1) defines the position of a bright or dark fringe, the term (2) defines the shift occurred in the particular fringe due to the introduction of a transparent plate.

Constructive and Destructive Interference

For constructive interference, the path difference must be an integral multiple of the wavelength.

Thus for a bright fringe to be at 'y',

$$n\lambda = y\, dD$$

Or, ynth = nλ Dd

Where n = ±0,1,2,3.....

The oth fringe represents the central bright fringe.

Similarly, the expression for a dark fringe in Young's double slit experiment can be found by setting the path difference as:

$$\Delta l = (n+12)\lambda$$

This simplifies to yn = (n+12)λ Dd

Note that these expressions require that θ be very small. Hence yD needs to be very small. This implies D should be very large and y should be small. This, in turn, requires that the formula works best for fringes close to the central maxima. In general, for best results, dD must be kept as small as possible for a good interference pattern.

The Young's double slit experiment was a watershed moment in scientific history because it firmly established that light indeed behaved as a wave.

The Double Slit Experiment was later conducted using electrons, and to everyone's surprise, the pattern generated was similar as expected with light. This would forever change our understanding of matter and particles, forcing us to accept that matter like light also behaves like a wave.

Diffraction

"The bending of light waves around the corners of an opening or obstacle and spreading of light waves into geometrical shadow is called diffraction. "Diffraction effect depends upon the size of obstacle .Diffraction of light takes place if the size of obstacle is comparable to the wavelength of light. Light waves are very small in wavelength, i.e, from 4×10^{-7} m to 7×10^{-7} m. If the size of opening or obstacle is near to this limit, only then we can observe the phenomenon of diffraction.

Types of Diffraction in Physics

Diffraction of light can be divided into two types:

- Fraunhofer Diffraction
- Fresnel Diffraction

Difference between Fresnel and Fraunhofer Diffraction

Fraunhofer Diffraction

In Fraunhofer diffraction:

- Source and the screen are far away from each other.
- Incident wave fronts on the diffracting obstacle are plane.
- Diffraction obstacle give rise to wave fronts which are also plane.
- Plane diffracting wave fronts are converged by means of a convex lens to produce diffraction pattern.

Fresnel Diffraction

In Fresnel diffraction:

- Source and screen are not far away from each other.
- Incident wave fronts are spherical.
- Wave fronts leaving the obstacles are also spherical.
- Convex lens is not needed to converge the spherical wave fronts.

Diffraction of Light

In Young's double slit experiment for the interference of light, the central region of the fringe system is bright. If light travels in straight path, the central region should appear dark i.e., the shadow of the screen between the two slits. Another simple experiment can be performed for exhibiting the same effect.

Consider that a small and smooth ball of about 3 mm in diameter is illuminated by a point source of light. The shadow of the object is received on a screen as shown in figure. The shadow of the spherical object is not completely dark but has a bright spot at its centre. According to Huygens's principle, each point on the rim of the sphere behaves as a source of secondary wavelets which illuminate the central region of the shadow.

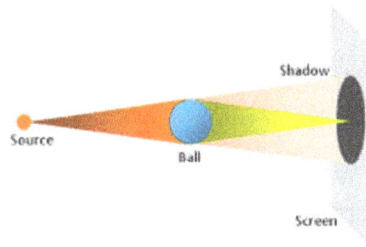

These two experiments clearly show that when light travels past an obstacle, it does not proceed exactly along a straight path, but bends around the obstacle. The phenomenon is found to be prominent when the wavelength of light is compared with the size of the obstacle or aperture of the slit. The diffraction of light occurs, in effect, due to the interference between rays coming from different parts of the same wave front.

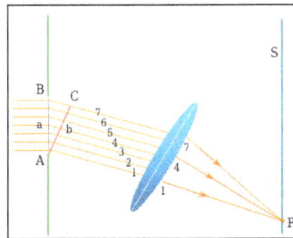

Diffraction due to Narrow Slit

Figure shows the experimental arrangement for studying diffraction of light due to narrow slit. The slit AB of width d is illuminated by a parallel beam of monochromatic light of wavelength λ. The screen S is placed parallel to the slit for observing the effects of the diffraction of light. A small portion of the incident wave front passes through the narrow slit. Each point of this topic of the wave front sends out secondary wavelets to the screen. These wavelets then interfere to produce the diffraction pattern. It becomes simple to deal with rays instead of wave fronts as shown in figure. In this figure, only nine rays have been drawn whereas actually there are a large number of them. Let us consider rays 1 and 5 which are in phase on in the wave front AB. When these reach the wave front AC, ray 5 would have a path difference ab say equal to $\lambda/2$. Thus,when these two rays reach point p on the screen; they will interfere destructively. Similarly, each pair 2 and 6,3 and 7,4 and 8 differ in path by $\lambda/2$ and will do the same. But the path difference ab=d/2 $\sin\theta$.

The equation for the first minimum is, then

$$d/2 \sin\theta = \lambda/2$$

or $d \sin\theta = \lambda$

In general,the conditions for different orders of minima on either side of centre are given by:

$$d \sin\theta = m\lambda$$

Where

$m = \pm (1,2,3,....)$

The region between any two consecutive minima both above and below O will be bright. A narrow slit, therefore, produces a series of bright and dark regions with the first bright region at the centre of the pattern.

Polarization

A light wave is an electromagnetic wave that travels through the vacuum of outer space. Light waves are produced by vibrating electric charges.an electromagnetic wave is a transverse wave that has both an electric and a magnetic component.

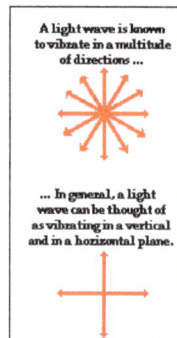

Let's suppose that we use the customary slinky to model the behavior of an electromagnetic wave. As an electromagnetic wave traveled towards you, then you would observe the vibrations of the slinky occurring in more than one plane of vibration. This is quite different than what you might notice if you were to look along a slinky and observe a slinky wave traveling towards you. Indeed, the coils of the slinky would be vibrating back and forth as the slinky approached; yet these vibrations would occur in a single plane of space. That is, the coils of the slinky might vibrate up and down or left and right. Yet regardless of their direction of vibration, they would be moving along the same linear direction as you sighted along the slinky. If a slinky wave were an electromagnetic wave, then the vibrations of the slinky would occur in multiple planes. Unlike a usual slinky wave, the electric and magnetic vibrations of an electromagnetic wave occur in numerous planes. A light wave that is vibrating in more than one plane is referred to as unpolarized light. Light emitted by the sun, by a lamp in the classroom, or by a candle flame is unpolarized light. Such light waves are created by electric charges that vibrate in a variety of directions, thus creating an electromagnetic wave that vibrates in a variety of directions. This concept of unpolarized light is rather difficult to visualize. In general, it is helpful to picture unpolarized light as a wave that has an average of half its vibrations in a horizontal plane and half of its vibrations in a vertical plane.

It is possible to transform unpolarized light into polarized light. Polarized light waves are light waves in which the vibrations occur in a single plane. The process of transforming unpolarized light into polarized light is known as polarization. There are a variety of methods of polarizing light. The four methods are discussed below.

Polarization by use of a Polaroid Filter

The most common method of polarization involves the use of a Polaroid filter. Polaroid filters are made of a special material that is capable of blocking one of the two planes of vibration of an electromagnetic wave. (Remember, the notion of two planes or directions of vibration is merely a simplification that helps us to visualize the wavelike nature of the electromagnetic wave.) In this sense, a Polaroid serves as a device that filters out one-half of the vibrations upon transmission of the light through the filter. When unpolarized light is transmitted through a Polaroid filter, it emerges with one-half the intensity and with vibrations in a single plane; it emerges as polarized light.

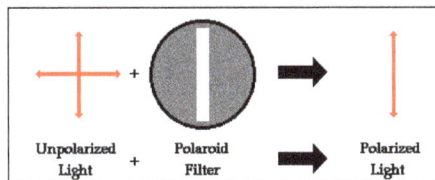

A Polaroid filter is able to polarize light because of the chemical composition of the filter material. The filter can be thought of as having long-chain molecules that are aligned within the filter in the same direction. During the fabrication of the filter, the long-chain molecules are stretched across the filter so that each molecule is (as much as possible) aligned in say the vertical direction. As unpolarized light strikes the filter, the portion of the waves vibrating in the vertical direction are absorbed by the filter. The general rule is that the electromagnetic vibrations that are in a direction parallel to the alignment of the molecules are absorbed.

The alignment of these molecules gives the filter a polarization axis. This polarization axis extends across the length of the filter and only allows vibrations of the electromagnetic wave that are parallel to the axis to pass through. Any vibrations that are perpendicular to the polarization axis are blocked by the filter. Thus, a Polaroid filter with its long-chain molecules aligned horizontally will have a polarization axis aligned vertically. Such a filter will block all horizontal vibrations and allow the vertical vibrations to be transmitted. On the other hand, a Polaroid filter with its long-chain molecules aligned vertically will have a polarization axis aligned horizontally; this filter will block all vertical vibrations and allow the horizontal vibrations to be transmitted.

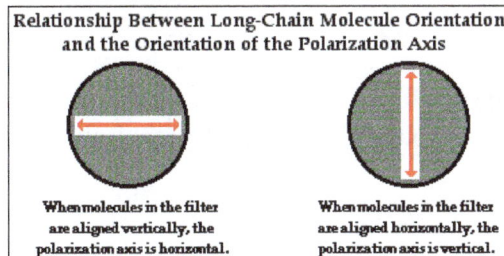

Relationship Between Long-Chain Molecule Orientation and the Orientation of the Polarization Axis

When molecules in the filter are aligned vertically, the polarization axis is horizontal.

When molecules in the filter are aligned horizontally, the polarization axis is vertical.

Polarization of light by use of a Polaroid filter is often demonstrated in a Physics class through a variety of demonstrations. Filters are used to look through and view objects. The filter does not distort the shape or dimensions of the object; it merely serves to produce a dimmer image of the object since one-half of the light is blocked as it passed through the filter. A pair of filters is often placed back to back in order to view objects looking through two filters. By slowly rotating the second filter, an orientation can be found in which all the light from an object is blocked and the object can no longer be seen when viewed through two filters. In this demonstration, the light was polarized upon passage through the first filter; perhaps only vertical vibrations were able to pass through. These vertical vibrations were then blocked by the second filter since its polarization filter is aligned in a horizontal direction. While you are unable to see the axes on the filter, you will know when the axes are aligned perpendicular to each other because with this orientation, all light is blocked. So by use of two filters, one can completely block all of the light that is incident upon the set; this will only occur if the polarization axes are rotated such that they are perpendicular to each other.

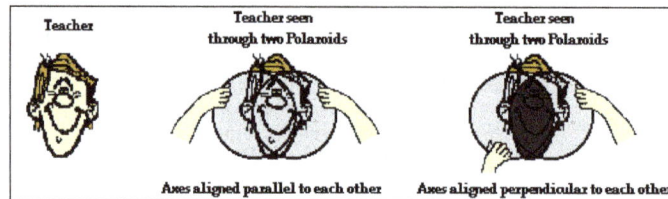

A picket-fence analogy is often used to explain how this dual-filter demonstration works. A picket fence can act as a polarizer by transforming an unpolarized wave in a rope into a wave that vibrates in a single plane. The spaces between the pickets of the fence will allow vibrations that are parallel to the spacings to pass through while blocking any vibrations that are perpendicular to the spacings. Obviously, a vertical vibration would not have the room to make it through a horizontal spacing. If two picket fences are oriented such that the pickets are both aligned vertically, then vertical vibrations will pass through both fences. On the other hand, if the pickets of the second fence are aligned horizontally, then the vertical vibrations that pass through the first fence will be blocked by the second fence. This is depicted in the diagram below.

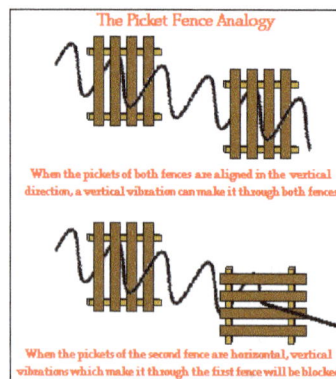

In the same manner, two Polaroid filters oriented with their polarization axes perpendicular to each other will block all the light. Now that's a pretty cool observation that could never be explained by a particle view of light.

Polarization by Reflection

Unpolarized light can also undergo polarization by reflection off of nonmetallic surfaces. The extent to which polarization occurs is dependent upon the angle at which the light approaches the surface and upon the material that the surface is made of. Metallic surfaces reflect light with a variety of vibrational directions; such reflected light is unpolarized. However, nonmetallic surfaces such as asphalt roadways, snowfields and water reflect light such that there is a large concentration of vibrations in a plane parallel to the reflecting surface. A person viewing objects by means of light reflected off of nonmetallic surfaces will often perceive a glare if the extent of polarization is large. Fishermen are familiar with this glare since it prevents them from seeing fish that lie below the water. Light reflected off a lake is partially polarized in a direction parallel to the water's surface. Fishermen know that the use of glare-reducing sunglasses with the proper polarization axis allows for the blocking of this partially polarized light. By blocking the plane-polarized light, the glare is reduced and the fisherman can more easily see fish located under the water.

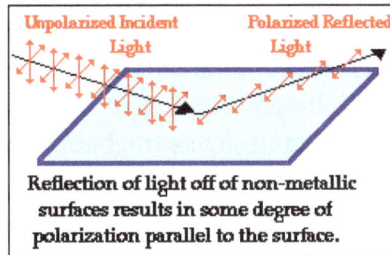

Reflection of light off of non-metallic surfaces results in some degree of polarization parallel to the surface.

Polarization by Refraction

Polarization can also occur by the refraction of light. Refraction occurs when a beam of light passes from one material into another material. At the surface of the two materials, the path of the beam changes its direction. The refracted beam acquires some degree of polarization. Most often, the polarization occurs in a plane perpendicular to the surface. The polarization of refracted light is often demonstrated in a Physics class using a unique crystal that serves as a double-refracting crystal. Iceland Spar, a rather rare form of the mineral calcite, refracts incident light into two different paths. The light is split into two beams upon entering the crystal. Subsequently, if an object is viewed by looking through an Iceland Spar crystal, two images will be seen. The two images are the result of the double refraction of light. Both refracted light beams are polarized - one in a direction parallel to the surface and the other in a direction perpendicular to the surface. Since these two refracted rays are polarized with a perpendicular orientation, a polarizing filter can be used to completely block one of the images. If the polarization axis of the filter is aligned perpendicular to the plane of polarized light, the light is completely blocked by the filter; meanwhile the second image is as bright as can be. And if the filter is then turned 90-degrees in either direction, the second image reappears and the first image disappears. Now that's pretty neat observation that could never be observed if light did not exhibit any wavelike behavior.

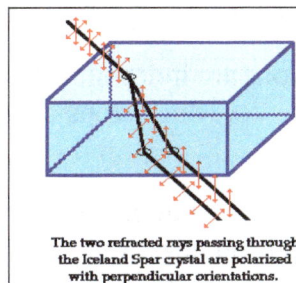

The two refracted rays passing through the Iceland Spar crystal are polarized with perpendicular orientations.

Polarization by Scattering

Polarization also occurs when light is scattered while traveling through a medium. When light strikes the atoms of a material, it will often set the electrons of those atoms into vibration. The vibrating electrons then produce their own electromagnetic wave that is radiated outward in all directions. This newly generated wave strikes neighboring atoms, forcing their electrons into vibrations at the same original frequency. These vibrating electrons produce another electromagnetic wave that is once more radiated outward in all directions. This absorption and reemission of light waves causes the light to be scattered about the medium. (This process of scattering contributes to the blueness of our skies, This scattered light is partially polarized. Polarization by scattering is observed as light passes through our atmosphere. The scattered light often produces a glare in the skies. Photographers know that this partial polarization of scattered light leads to photographs characterized by a washed-out sky. The problem can easily be corrected by the use of a Polaroid filter. As the filter is rotated, the partially polarized light is blocked and the glare is reduced. The photographic secret of capturing a vivid blue sky as the backdrop of a beautiful foreground lies in the physics of polarization and Polaroid filters.

Particle Nature of Light

The Photoelectric Effect

The first evidence for the particle nature of light was obtained by in 1887 by Hertz in his experiments on electromagnetic radiation. He noticed that when ultra-violet radiation fell on a metal electrode, an electric charge was produced. In 1900, by deflecting the emitted charges in a magnetic field, Lenard showed that the charges were electrons. This effect is called the photoelectric effect and the emitted photons are called photoelectrons.

A schematic diagram of the setup of an experiment for studying the photoelectric effect is shown below.

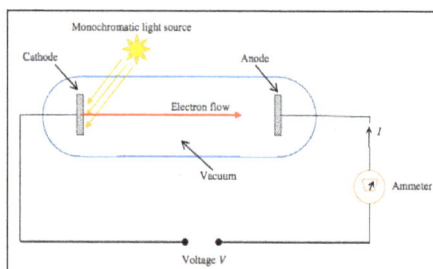

The important results from the experiment are:

1. The kinetic energy of the electrons are independent of the light intensity. This is shown by applying a stopping potential, $-V_0$, which stops all photoelectrons no matter how strong the light is.

2. For fixed light intensity, there is a maximum possible photocurrent, which is reached when the applied voltage is increased from negative to positive values. The maximum photocurrent is proportional to the light intensity.

3. The maximum kinetic energy of the photoelectrons depends on the frequency of the light but not on its intensity (i.e. the stopping voltage depends on the frequency but not the intensity of light).

4. There is a threshold frequency below which no photoelectrons are emitted. The threshold frequency is different for different metals.

These results are at odds with the expectations of classical physics. In the classical picture, the energy of a light wave increases with the intensity of the light. Therefore, the photoelectrons should get more kinetic energy as the light intensity increases. In contrast, the experiments show that the maximum kinetic energy of the photoelectrons is independent of the light intensity but depends on the light frequency. The existence of a threshold frequency is inexplicable in classical theory.

Einstein's Theory

Building on Planck's explanation of thermal radiation, Einstein developed a quantum theory for the photoelectric effect. Einstein suggested that electromagnetic radiation is quantized, i.e. the energy of a light wave is not continuously distributed in space but consists of a number of discrete energy quanta which are localized at points in space. These quanta, which we now call photons, move without dividing and can only be emitted or absorbed as whole units. Einstein proposed that the energy of a photon is related to the frequency of light by,

$$E = hf,$$

where h is Planck's constant. Hence Einstein is proposing that light has simultaneously particle-like and wave-like properties (often referred to as wave – particle duality). Some experiments reveal the wave-like properties (interference effects) and others the particle-like properties (photoelectric effect, pair creation).

To explain the photoelectric effect, Einstein proposed that a single photon gives up its entire energy in liberating a single electron from the metal. In leaving the metal the electron gives up some of its energy to other electrons. The remaining energy appears as the kinetic energy of the electron. Conservation of energy requires,

$$hf = \phi + \frac{1}{2}mv_{max}^2,$$

Where ϕ is the energy needed to leave the metal and is called the work function. The threshold frequency is such that,

$$hf_{threshold} = \phi$$

Typical values for the work function are 2 – 6 eV, and corresponding values for the threshold frequency are in the visible to near UV part of the electromagnetic spectrum.

Photon Momentum

Since photons are massless, their energy and momentum are related by,

$$E = cp.$$

Using equation $E = hf$, we see that,

$$P = \frac{E}{c} = \frac{hf}{c} = \frac{h}{\lambda}$$

Where λ is the wavelength of the light.

Relations using Angular Frequency and Wave Number

The angular frequency is,

$$\omega = 2\pi f \ ,$$

And the wave number is,

$$k = \frac{2\pi}{\lambda}.$$

In terms of these new quantities, the energy and momentum of a photon are,

$$E = \hbar\omega,$$

And

$$p = \hbar k,$$

Where the reduced Planck's constant, $\hbar$, is related to Planck's constant, h, by

$$\hbar = \frac{h}{2\pi}.$$

Reference

- What-is-light: oxinst.com, Retrieved 3 May, 2019
- Optics, science: britannica.com, Retrieved 23 February, 2019
- Geometrical-optics, what-is-optics: optics4kids.org, Retrieved 15 July, 2019
- Introduction-wave-optics, wave-optics, physics: emedicalprep.com, Retrieved 18 March, 2019
- Huygens-principle, science: britannica.com, Retrieved 6 January, 2019
- Huygens-principle, wave-optics, physics: toppr.com, Retrieved 16 August, 2019
- Interference-of-light, pure-sciences: scienceabc.com, Retrieved 26 June, 2019
- Interference-of-light-waves-and-youngs-experiment, wave-optics, physics: toppr.com, Retrieved 28 February, 2019
- Youngs-double-slit-experiment: byjus.com, Retrieved 8 April, 2019
- Diffraction-of-light: physicsabout.com, Retrieved 19 January, 2019

7

Special Relativity

Special relativity is a physical theory that explains the relationship between space and time. It is based on two postulates i.e. in all inertial frames of reference, the laws of physics are identical and the speed of light in a vacuum is the same for all observers. All the diverse principles of special relativity have been carefully analysed in this chapter.

Special relativity is a fundamental physics theory about space and time that was developed by Albert Einstein in 1905 as a modification of Newtonian physics. It was created to deal with some pressing theoretical and experimental issues in the physics of the time involving light and electrodynamics. The predictions of special relativity correspond closely to those of Newtonian physics at speeds which are low in comparison to that of light, but diverge rapidly for speeds which are a significant fraction of the speed of light. Special relativity has been experimentally tested on numerous occasions since its inception, and its predictions have been verified by those tests.

Albert Einstein during a lecture

Einstein postulated that the speed of light is the same for all observers, irrespective of their motion relative to the light source. This was in total contradiction to classical mechanics, which had been accepted for centuries. Einstein's approach was based on thought experiments and calculations. In 1908, Hermann Minkowski reformulated the theory based on different postulates of a more geometrical nature. His approach depended on the existence of certain interrelations between space and time, which were considered completely separate in classical physics. This reformulation set the stage for further developments of physics.

Special relativity makes numerous predictions that are incompatible with Newtonian physics (and everyday intuition). The first such prediction described by Einstein is called the relativity of simultaneity, under which observers who are in motion with respect to each other may disagree on whether two events occurred at the same time or one occurred before the other. The other major predictions of special relativity are time dilation (under which a moving clock ticks more slowly than when it is at rest with respect the observer), length contraction (under which a moving rod may be found to be shorter than when it is at rest with respect to the observer), and the equivalence of mass and energy (written as $E=mc^2$). Special relativity predicts a non-linear velocity addition formula, which prevents speeds greater than that of light from being observed. Special relativity also explains why Maxwell's equations of electromagnetism are correct in any frame of reference, and how an electric field and a magnetic field are two aspects of the same thing.

Special relativity has received experimental support in many ways, and it has been proven far more accurate than Newtonian mechanics. The most famous experimental support is the Michelson-Morley experiment, the results of which (showing that the speed of light is a constant) was one factor that motivated the formulation of the theory of special relativity. Other significant tests are the Fizeau experiment (which was first done decades before special relativity was proposed), the detection of the transverse Doppler effect, and the Haefele-Keating experiment. Today, scientists are so comfortable with the idea that the speed of light is always the same that the meter is now defined as being the distance traveled by light in 1/299,792,458th of a second. This means that the speed of light is now defined as being 299,792,458 m/s.

Reference Frames and Galilean Relativity

A reference frame is simply a selection of what constitutes stationary objects. Once the velocity of a certain object is arbitrarily defined to be zero, the velocity of everything else in the universe can be measured relative to it. When a train is moving at a constant velocity past a platform, one may either say that the platform is at rest and the train is moving or that the train is at rest and the platform is moving past it. These two descriptions correspond to two different reference frames. They are respectively called the rest frame of the platform and the rest frame of the train (sometimes simply the platform frame and the train frame).

The question naturally arises, can different reference frames be physically differentiated? In other words, can one conduct some experiments to claim that "we are now in an absolutely stationary reference frame?" Aristotle thought that all objects tend to cease moving and become at rest if there were no forces acting on them. Galileo challenged this idea and argued that the concept of absolute motion was unreal. All motion was relative. An observer who couldn't refer to some isolated object (if, say, he was imprisoned inside a closed spaceship) could never distinguish whether according to some external observer he was at rest or moving with constant velocity. Any experiment he could conduct would give the same result in both cases. However, accelerated reference frames are experimentally distinguishable. For example, if an astronaut moving in free space saw that the tea in his tea-cup was slanted rather than horizontal, he would be able to infer that his spaceship was accelerated. Thus not all reference frames are equivalent, but people have a class of reference frames, all moving at uniform velocity with respect to each other, in all of which Newton's first law holds. These are called the inertial reference frames and are fundamental to both classical mechanics and SR. Galilean relativity thus states that the laws of physics can not depend on absolute velocity, they must stay the same in any inertial reference frame. Galilean relativity is thus a fundamental principle in classical physics.

Mathematically, it says that if one transforms all velocities to a different reference frame, the laws of physics must be unchanged. What is this transformation that must be applied to the velocities? Galileo gave the common-sense "formula" for adding velocities: If

1. Particle P is moving at velocity v with respect to reference frame A and

2. Reference frame A is moving at velocity u with respect to reference frame B, then

3. The velocity of P with respect to B is given by v + u.

The formula for transforming coordinates between different reference frames is called the Galilean transformation. The principle of Galilean relativity then demands that laws of physics be unchanged if the Galilean transformation is applied to them. Laws of classical mechanics, like Newton's second law, obey this principle because they have the same form after applying the transformation. As Newton's law involves the derivative of velocity, any constant velocity added in a Galilean transformation to a different reference frame contributes nothing (the derivative of a constant is zero). Addition of a time-varying velocity (corresponding to an accelerated reference frame) will however change the formula, since Galilean relativity only applies to non-accelerated inertial reference frames.

Time is the same in all reference frames because it is absolute in classical mechanics. All observers measure exactly the same intervals of time and there is such a thing as an absolutely correct clock.

Invariance of Length

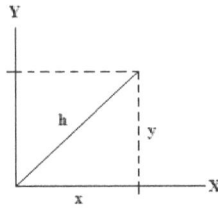

Pythagoras theorem

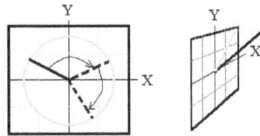

The length of an object is constant on the plane during rotations
on the plane but not during rotations out of the plane.

In special relativity, space and time are joined into a unified four-dimensional continuum called spacetime. To gain a sense of what spacetime is like, we must first look at the Euclidean space of Newtonian physics.

This approach to the theory of special relativity begins with the concept of "length." In everyday experience, it seems that the length of objects remains the same no matter how they are rotated or moved from place to place; as a result the simple length of an object doesn't appear to change or is "invariant." However, as is shown in the illustrations below, what is actually being suggested is that length seems to be invariant in a three-dimensional coordinate system.

The length of a line in a two-dimensional Cartesian coordinate system is given by Pythagoras' theorem:

$$h^2 = x^2 + y^2$$

One of the basic theorems of vector algebra is that the length of a vector does not change when it is rotated. However, a closer inspection tells us that this is only true if we consider rotations confined to the plane. If we introduce rotation in the third dimension, then we can tilt the line out of the plane. In this case the projection of the line on the plane will get shorter. Does this mean length is not invariant? Obviously not. The world is three-dimensional and in a 3D Cartesian coordinate system the length is given by the three-dimensional version of Pythagoras's theorem:

$$k^2 = x^2 + y^2 + z^2.$$

This is invariant under all rotations. The apparent violation of invariance of length only happened because we were "missing" a dimension. It seems that, provided all the directions in which an object can be tilted or arranged are represented within a coordinate system, the length of an object does not change under rotations. A 3-dimensional coordinate system is enough in classical mechanics because time is assumed absolute and independent of space in that context. It can be considered separately.

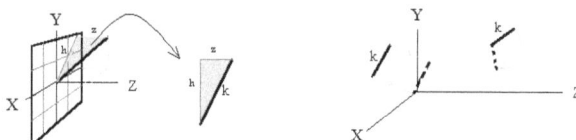

Invariance in a 3D coordinate system: Pythagoras theorem gives $k^2 = h^2 + z^2$ but $h^2 = x^2 + y^2$ therefore $k^2 = x^2 + y^2 + z^2$. The length of an object is constant whether it is rotated or moved from one place to another in a 3D coordinate system.

Note that invariance of length is not ordinarily considered a dynamic principle, not even a theorem. It is simply a statement about the fundamental nature of space itself. Space as we ordinarily conceive it is called a three-dimensional Euclidean space, because its geometrical structure is described by the principles of Euclidean geometry. The formula for distance between two points is a fundamental property of an Euclidean space, it is called the Euclidean metric tensor (or simply the Euclidean metric). In general, distance formulas are called metric tensors.

Note that rotations are fundamentally related to the concept of length. In fact, one may define length or distance to be that which stays the same (is invariant) under rotations, or define rotations to be that which keep the length invariant. Given any one, it is possible to find the other. If we know the distance formula, we can find out the formula for transforming coordinates in a rotation. If, on the other hand, we have the formula for rotations then we can find out the distance formula.

The Postulates of Special Relativity

Einstein developed Special Relativity on the basis of two postulates:

- First postulate—Special principle of relativity—The laws of physics are the same in all inertial frames of reference. In other words, there are no privileged inertial frames of reference.

- Second postulate—Invariance of c—The speed of light in a vacuum is independent of the motion of the light source.

Special Relativity can be derived from these postulates, as was done by Einstein in 1905. Einstein's postulates are still applicable in the modern theory but the origin of the postulates is more explicit. It was shown above how the existence of a universally constant velocity (the speed of light) is a consequence of modeling the universe as a particular four dimensional space having certain specific properties. The principle of relativity is a result of Minkowski structure being preserved under Lorentz transformations, which are postulated to be the physical transformations of inertial reference frames.

The Minkowski Formulation

Hermann Minkowski

After Einstein derived special relativity formally from the counterintuitive proposition that the speed of light is the same to all observers, the need was felt for a more satisfactory formulation. Minkowski, building on mathematical approaches used in non-Euclidean geometry and the mathematical work of Lorentz and Poincaré, realized that a geometric approach was the key. Minkowski showed in 1908 that Einstein's new theory could be explained in a natural way if the concept of separate space and time is replaced with one four-dimensional continuum called spacetime. This was a groundbreaking concept, and Roger Penrose has said that relativity was not truly complete until Minkowski reformulated Einstein's work.

The concept of a four-dimensional space is hard to visualize. It may help at the beginning to think simply in terms of coordinates. In three-dimensional space, one needs three real numbers to refer to a point. In the Minkowski space, one needs four real numbers (three space coordinates and one time coordinate) to refer to a point at a particular instant of time. This point at a particular instant of time, specified by the four coordinates, is called an event. The distance between two different events is called the spacetime interval.

A path through the four-dimensional spacetime, usually called Minkowski space, is called a world line. Since it specifies both position and time, a particle having a known world line has a completely determined trajectory and velocity. This is just like graphing the displacement of a particle moving in a straight line against the time elapsed. The curve contains the complete motional information of the particle.

In the same way as the measurement of distance in 3D space needed all three coordinates we must include time as well as the three space coordinates when calculating the distance in Minkowski space (henceforth called M). In a sense, the spacetime interval provides a combined estimate of how far two events occur in space as well as the time that elapses between their occurrence.

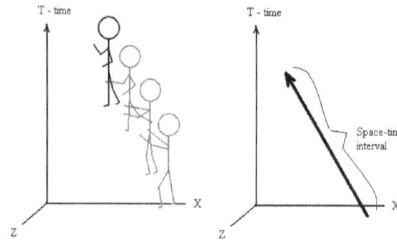

The spacetime interval.

But there is a problem. Time is related to the space coordinates, but they are not equivalent. Pythagoras's theorem treats all coordinates on an equal footing. We can exchange two space coordinates without changing the length, but we can not simply exchange a space coordinate with time, they are fundamentally different. It is an entirely different thing for two events to be separated in space and to be separated in time. Minkowski proposed that the formula for distance needed a change. He found that the correct formula was actually quite simple, differing only by a sign from Pythagoras's theorem:

$$s^2 = x^2 + y^2 + z^2 - \left(ct\right)^2$$

where c is a constant and t is the time coordinate. Multiplication by c, which has the dimension ms^{-1}, converts the time to units of length and this constant has the same value as the speed of light. So the spacetime interval between two distinct events is given by

$$s^2 = (x_2 - x_1)^2 + (y_2 - y_1)^2 + (z_2 - z_1)^2 + (t_2 - t_1)^2$$

There are two major points to be noted. Firstly, time is being measured in the same units as length by multiplying it by a constant conversion factor. Secondly, and more importantly, the time-coordinate has a different sign than the space coordinates. This means that in the four-dimensional spacetime, one coordinate is different from the others and influences the distance differently. This new 'distance' may be zero or even negative. This new distance formula, called the metric of the spacetime, is at the heart of relativity. This distance formula is called the metric tensor of M. This minus sign means that a lot of our intuition about distances can not be directly carried over into spacetime intervals. For example, the spacetime interval between two events separated both in time and space may be zero. From now on, the terms distance formula and metric tensor will be used interchangeably, as will be the terms Minkowski metric and spacetime interval.

In Minkowski spacetime the spacetime interval is the invariant length, the ordinary 3D length is not required to be invariant. The spacetime interval must stay the same under rotations, but ordinary lengths can change. Just like before, we were missing a dimension. Note that everything this far are merely definitions. We define a four-dimensional mathematical construct which has a special formula for distance, where distance means that which stays the same under rotations (alternatively, one may define a rotation to be that which keeps the distance unchanged).

Now comes the physical part. Rotations in Minkowski space have a different interpretation than ordinary rotations. These rotations correspond to transformations of reference frames. Passing from one reference frame to another corresponds to rotating the Minkowski space. An intuitive justification for this is given below, but mathematically this is a dynamical postulate just like assuming that physical laws must stay the same under Galilean transformations (which seems so intuitive that we don't usually recognize it to be a postulate).

Since by definition rotations must keep the distance same, passing to a different reference frame must keep the spacetime interval between two events unchanged. This requirement can be used to derive an explicit mathematical form for the transformation that must be applied to the laws of physics (compare with the application of Galilean transformations to classical laws) when shifting reference frames. These transformations are called the Lorentz transformations. Just like the Galilean transformations are the mathematical statement of the principle of Galilean relativity in classical mechanics, the Lorentz transformations are the mathematical form of Einstein's principle of relativity. Laws of physics must stay the same under Lorentz transformations. Maxwell's equations and Dirac's equation satisfy this property, and hence, they are relativistically correct laws (but classically incorrect, since they don't transform correctly under Galilean transformations).

With the statement of the Minkowski metric, the common name for the distance formula given above, the theoretical foundation of special relativity is complete. The entire basis for special relativity can be summed up by the geometric statement "changes of reference frame correspond to rotations in the 4D Minkowski spacetime, which is defined to have the distance formula given above." The unique dynamical predictions of SR stem from this geometrical property of spacetime. Special relativity may be said to be the physics of Minkowski spacetime. In this case of spacetime, there are six independent rotations to be considered. Three of them are the standard rotations on a plane in two directions of space. The other three are rotations in a plane of both space and time: These rotations correspond to a change of velocity, and are described by the traditional Lorentz transformations.

As has been mentioned before, one can replace distance formulas with rotation formulas. Instead of starting with the invariance of the Minkowski metric as the fundamental property of spacetime, one may state (as was done in classical physics with Galilean relativity) the mathematical form of the Lorentz transformations and require that physical laws be invariant under these transformations. This makes no reference to the geometry of spacetime, but will produce the same result. This was in fact the traditional approach to SR, used originally by Einstein himself. However, this approach is often considered to offer less insight and be more cumbersome than the more natural Minkowski formalism.

Reference Frames and Lorentz Transformations

In classical mechanics coordinate frame changes correspond to Galilean transfomations of the coordinates. Is this adequate in the relativistic Minkowski picture?

Suppose there are two people, Bill and John, on separate planets that are moving away from each other. Bill and John are on separate planets so they both think that they are stationary. John draws a graph of Bill's motion through space and time and this is shown in the illustration below:

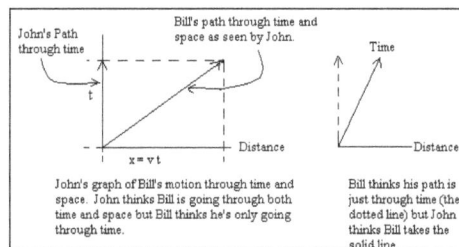

John's view of Bill and Bill's view of himself

John sees that Bill is moving through space as well as time but Bill thinks he is moving through time alone. Bill would draw the same conclusion about John's motion. In fact, these two views, which would be classically considered a difference in reference frames, are related simply by a co-ordinate transformation in M. Bill's view of his own world line and John's view of Bill's world line are related to each other simply by a rotation of coordinates. One can be transformed into the other by a rotation of the time axis. Minkowski geometry handles transformations of reference frames in a very natural way.

Changes in reference frame, represented by velocity transformations in classical mechanics, are represented by rotations in Minkowski space. These rotations are called Lorentz transformations. They are different from the Galilean transformations because of the unique form of the Minkowski metric. The Lorentz transformations are the relativistic equivalent of Galilean transformations. Laws of physics, in order to be relativistically correct, must stay the same under Lorentz transformations. The physical statement that they must be same in all inertial reference frames remains unchanged, but the mathematical transformation between different reference frames changes. Newton's laws of motion are invariant under Galilean rather than Lorentz transformations, so they are immediately recognizable as non-relativistic laws and must be discarded in relativistic physics. Schrödinger's equation is also non-relativistic.

Maxwell's equations are trickier. They are written using vectors and at first glance appear to transform correctly under Galilean transformations. But on closer inspection, several questions are apparent that can not be satisfactorily resolved within classical mechanics. They are indeed invariant under Lorentz transformations and are relativistic, even though they were formulated before the discovery of special relativity. Classical electrodynamics can be said to be the first relativistic theory in physics. To make the relativistic character of equations apparent, they are written using 4-component vector like quantities called 4-vectors. 4-Vectors transform correctly under Lorentz transformations. Equations written using 4-vectors are automatically relativistic. This is called the manifestly covariant form of equations. 4-Vectors form a very important part of the formalism of special relativity.

Einstein's Postulate: The Constancy of the Speed of Light

Einstein's postulate that the speed of light is a constant comes out as a natural consequence of the Minkowski formulation

Proposition 1:

When an object is traveling at c in a certain reference frame, the spacetime interval is zero.

Proof:

The spacetime interval between the origin-event $(0,0,0,0)$ and an event (x, y,z, t) is

$$s^2 = x^2 + y^2 + z^2 - (ct)^2.$$

The distance travelled by an object moving at velocity v for t seconds is:

$$\sqrt{x^2 + y^2 + z^2} = vt$$

giving

$$s^2 = (vt)^2 - (ct)^2$$

Since the velocity v equals c we have

$$s^2 = (ct)^2 - (ct)^2$$

Hence the spacetime interval between the events of departure and arrival is given by

$$s^2 = 0$$

Proposition 2:

An object traveling at c in one reference frame is traveling at c in all reference frames.

Proof:

Let the object move with velocity v when observed from a different reference frame. A change in reference frame corresponds to a rotation in M. Since the spacetime interval must be conserved under rotation, the spacetime interval must be the same in all reference frames. In proposition 1 we showed it to be zero in one reference frame, hence it must be zero in all other reference frames. We get that

$$(vt)^2 - (ct)^2 = 0$$

which implies

$$|v| = c.$$

The paths of light rays have a zero spacetime interval, and hence all observers will obtain the same value for the speed of light. Therefore, when assuming that the universe has four dimensions that are related by Minkowski's formula, the speed of light appears as a constant, and does not need to be assumed (postulated) to be constant as in Einstein's original approach to special relativity.

Length Contraction

The first of the interesting consequences of the Lorentz Transformation is that length no longer has an absolute meaning: the length of an object depends on its motion relative to the frame of reference in which its length is being measured. Let us consider a rod moving with a velocity vx relative to a frame of reference S , and lying along the X axis. This rod is then stationary relative to a frame of reference S' which is also moving with a velocity v_x relative to S .

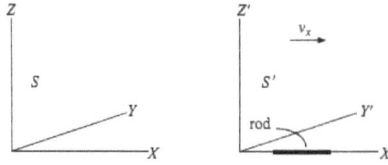

Figure: A rod of length at rest in reference frame S' which is moving with a velocity v_x with respect to another frame S .

As the rod is stationary in S' , the ends of the rod will have coordinates x'_1 and x'_2 which remain fixed as functions of the time in S' . The length of the rod, as measured in S' is then

$$l_0 = x'_2 - x'_1$$

where l_0 is known as the proper length of the rod i.e. l_0 is its length as measured in a frame of reference in which the rod is stationary. Now suppose that we want to measure the length of the rod as measured with respect to S . In order to do this, we measure the X coordinates of the two ends of the rod at the same time t, as measured by the clocks in S . Let x_2 and x_1 be the X coordinates of the two ends of the rod as measured in S at this time t. The two clocks positioned at x_2 and x_1 both read t when the two ends of the rod coincided with the points x_2 and x_1. Turning now to the Lorentz Transformation equations, we see that we must have

$$\left. \begin{aligned} x'_1 &= \gamma(x_1 - v_x t) \\ x'_2 &= \gamma(x_2 - v_x t). \end{aligned} \right\}$$

We then define the length of the rod as measured in the frame of reference S to be

$$l = x_2 - x_1$$

where the important point to be re-emphasized is that this length is defined in terms of the positions of the ends of the rods as measured at the same time t in S. Using the above equations we find,

$$l_0 = x'_2 - x'_1 = \gamma(x_2 - x_1) = \gamma l$$

which gives for l

$$l = \gamma^{-1} l_0 = \sqrt{1 - (v_x / c)^2} \, l_0.$$

But for $v_x < c$

$$\sqrt{1 - (v_x / c)^2} < 1$$

so that

$$l < l_0$$

Thus the length of the rod as measured in the frame of reference S with respect to which the rod is moving is shorter than the length as measured from a frame of reference S o relative to which the rod is stationary. A rod will be observed to have its maximum length when it is stationary in a frame of reference. The length so-measured, l_0 is known as its proper length.

This phenomenon is known as the Lorentz-Fitzgerald contraction. It is not the consequence of some force 'squeezing' the rod, but it is a real physical phenomenon with observable physical effects. Note however that someone who actually looks at this rod as it passes by will not see a shorter rod. If the time that is required for the light from each point on the rod to reach the observer's eye is taken into account, the overall effect is that of making the rod appear as if it is rotated in space.

Time Dilation

Time dilation, in the theory of special relativity is the "slowing down" of a clock as determined by an observer who is in relative motion with respect to that clock. In special relativity, an observer in inertial (i.e., nonaccelerating) motion has a well-defined means of determining which events occur simultaneously with a given event. A second inertial observer, who is in relative motion with respect to the first, however, will disagree with the first observer regarding which events are simultaneous with that given event. (Neither observer is wrong in this determination; rather, their disagreement merely reflects the fact that simultaneity is an observer-dependent notion in special relativity.) A notion of simultaneity is required in order to make a comparison of the rates of clocks carried by the two observers. If the first observer's notion of simultaneity is used, it is found that the second observer's clock runs slower than the first observer's by a factor of Square root of $\sqrt{1-\left(v^2/c^2\right)}$, where v is the relative velocity of the observers and c equals 299,792 km (186,282 miles) per second—i.e., the speed of light. Similarly, using the second observer's notion of simultaneity, it is found that the first observer's clock runs slower by the same factor. Thus, each inertial observer determines that all clocks in motion relative to that observer run slower than that observer's own clock.

A closely related phenomenon predicted by special relativity is the so-called twin paradox. Suppose one of two twins carrying a clock departs on a rocket ship from the other twin, an inertial observer, at a certain time, and they rejoin at a later time. In accordance with the time-dilation effect, the elapsed time on the clock of the twin on the rocket ship will be smaller than that of the inertial observer twin—i.e., the non-inertial twin will have aged less than the inertial observer twin when they rejoin.

The time-dilation effect predicted by special relativity has been accurately confirmed by observations of the increased lifetime of unstable elementary particles traveling at nearly the speed of light. The clock paradox effect also has been substantiated by experiments comparing the elapsed time of an atomic clock on Earth with that of an atomic clock flown in an airplane.

Relativity of Simultaneity

Another consequence of the transformation law for time is that events which occur simultaneously in one frame of reference will not in general occur simultaneously in any other frame of reference.

Thus, consider two events 1 and 2 which are simultaneous in S i.e. $t_1 = t_2$, but which occur at two different places x_1 and x_2. Then, in S ', the time interval between these two events is

$$t'_2 - t'_1 = \gamma\left(t_2 - v_x x_2 / c^2\right) - \gamma\left(t_1 - v_x x_1 / c^2\right)$$
$$= \gamma\left(x_1 - x_2\right)v_x / c^2$$
$$\neq 0 \ as \ x_1 \neq x_2.$$

Here t'_1 is the time registered on the clock in S' which coincides with the position x_1 in S at the instant t_1 that the event 1 occurs and similarly for t'_2 . Thus events which appear simultaneous in S are not simultaneous in S'. In fact the order in which the two events 1 and 2 are found to occur in will depend on the sign of $x_1 - x_2$ or v_x. It is only when the two events occur at the same point (i.e. $x_1 = x_2$) that the events will occur simultaneously in all frames of reference.

Relativistic Momentum

According to the first postulates of Einstein special theory of relativity, laws of physics are invariant in all inertial frames. We know that conservation of linear momentum in the absence of external force is the most fundamental and powerful law of physics. As per classical physics, the linear momentum of particle of mass 'm' moving with velocity u is defined as $p = mu$. We will see that this form of definition of momentum fails to obey the principle of conservation under relativistic dynamics. Let us consider two spherical clays A & B of equal masses 'm' move with velocities $u_A = V, u_B = -V$ and undergo inelastic collision with final velocities $V_A = V_B = 0$ measured from the rest frame s as shown in figure.

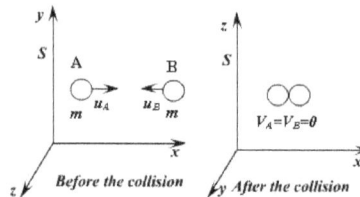

Figure: Inelastic collisions between two spherical clays A and B according to rest frame S .

Initial momentum, $pi = mu_A + mu_B = m(v - v) = 0$

Final momentum, $p_J = mV_A + mV_B = m \times 0 = 0$

Thus according to S frame conservation of momentum is obeyed.

Let us view the same problem with respect to S' frame moving with velocity v along x' direction as shown in figure.

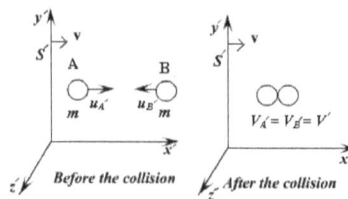

Figure: Inelastic collision between two spherical clays A & B as per inertial frame S'

The velocity transformation based on Lorentz transformation equations can be written as,

$$u'_x = \frac{u_x - v}{\left(1 - \dfrac{u_x v}{c^2}\right)} \; so$$

so

$$u'_A = \frac{u_A - v}{1 - \dfrac{u_A v}{c^2}} = \frac{v - v}{1 - \dfrac{v^2}{c^2}} = 0$$

$$u'_B = \frac{u_B - v}{1 - \dfrac{u_B v}{c^2}} = \frac{-v - v}{1 + \dfrac{v^2}{c^2}} = \frac{-2v}{1 + \dfrac{v^2}{c^2}}$$

$$V'_A = V'_B = V' = \frac{V_A - v}{1 - \dfrac{V_A v}{c^2}} = \frac{0 - v}{1 - 0} = -v$$

The total initial momentum in S' frame, $p_i' = \dfrac{-2mv}{1 + v^2/c^2}$

And the final momentum, $p_f' = -mv$

$$p_i' \neq p_f'$$

As per above equation, the momentum is not conserved in S' frame. We get different results with change in internal frame and it violates Einstein theory of relativity that laws of physics are invariant in different inertial frames. We have used Lorentz transformation for the transfer of velocity from S to S' frame. Lorentz theory is correct because it explains the cosntancy of velocity of light. So, the problem could be, the definition of linear momentum in relativistic dynamics. One has to take the appropriate definition of linear momentum such that it follows the fundamental laws physics like conservation of momentum in the absence of applied force. It should also reduce to the definition of classical physics, when the velocity of the body $u \langle\langle c$ [much small as compared to the velocity of light].

The relativistic momentum can be defined as,

$$p = m\frac{dx}{dt'}$$

where dx is the distance traveled by a particle of mass m and the time interval dt'. Here dt' is the proper time, that is the time measured with respect to the moving frame of velocity attached to the body.

As per the time dilaton relation, we know that,

$$dt = \frac{dt'}{\sqrt{1 - \dfrac{v^2}{c^2}}}$$

$$so, \; p = m\frac{dx}{dt} \cdot \frac{1}{\sqrt{1 - \dfrac{v^2}{c^2}}}$$

$$or \; p = \frac{mv}{\sqrt{1 - \dfrac{v^2}{c^2}}}$$

We can rewrite the above equation by taking mass m as m_0, where m_0 represents rest mass or mass of the particles as per Newtonian non-relativistic mechanics.

$$p = \frac{m_0 v}{\sqrt{1 - \dfrac{v^2}{c^2}}}$$

Relativistic Energy

We have seen that there is a requirement of definition of relativistic linear momentum and similarly one can get the expression of relativistic energy. We can make use of Newton's work-energy theorem. The work done, W on a particle of mass m_0 by an external force 'F' gives rise to gain in its kinetic energy.

$$dW = \overline{F}.d\overline{x}$$

$$or \; dW = \frac{\overline{dp}}{dt}.d\overline{x}$$

Change in kinetic energy

$$k_2 - k_1 = \int_1^2 dW = \int_1^2 \frac{\overline{dp}}{dt} \, \overline{dx}$$

By substituting the relativistic expression for $\overline{p}$ and by writing $\overline{dx} = \overline{v}dt$.

$$k_2 - k_1 = \int_1^2 \frac{d}{dt} \left[\frac{m_0 \overline{v}}{\sqrt{1 - \dfrac{v^2}{c^2}}} \right] \cdot \overline{v}dt$$

$$or \; k_2 - k_1 = \int_1^2 \overline{v} \cdot d\left[\frac{m_0 \overline{v}}{\sqrt{1 - \dfrac{v^2}{c^2}}} \right]$$

We know that,

$$d\left[\overline{v}\cdot\overline{p}\right] = \overline{v}\cdot\overline{dp} + \overline{p}\cdot\overline{dv}$$

$$or \quad \overline{v}\cdot\overline{dp} = d\left[\overline{v}\cdot\overline{p}\right] - \overline{p}\cdot\overline{dv}$$

So, equation before the above one can be written as,

$$k_2 - k_1 = \int_1^2 d\left[\overline{v}\cdot\frac{m_0\overline{v}}{\sqrt{1-\dfrac{v^2}{c^2}}}\right] - \int_1^2 \frac{m_0\overline{v}}{\sqrt{1-\dfrac{v^2}{c^2}}}\cdot\overline{dv}$$

$$= \frac{m_0 v^2}{\sqrt{1-\dfrac{v^2}{c^2}}}\Bigg|_0^{v} - \int_0^{v} \frac{m_0 v dv}{\sqrt{1-\dfrac{v^2}{c^2}}}$$

where it is assumed that the particle was at rest initially and its final velocities is $\overline{v}$

The second integration in above equation can be carried out easily based on substitution method,

Say, $1-\dfrac{v^2}{c^2} = \omega^2, -2\dfrac{v}{c^2}\,dv = 2\omega d\omega$

so, $\displaystyle\int_0^{v} \frac{m_0 v dv}{\sqrt{1-\dfrac{v^2}{c^2}}} = \int_0^{v} \frac{m_0(-d\omega)c^2}{\omega} = -m_0 c^2\ \omega\Big|_0^{v}$

$$= -m_0 c^2\left[\sqrt{1-\frac{v^2}{c^2}}\right]_0^{v}$$

$$= +m_0 c^2 - m_0 c^2\sqrt{1-\frac{v^2}{c^2}}$$

so,

$$k_2 - k_1 = K = \frac{m_0 v^2}{\sqrt{1-\dfrac{v^2}{c^2}}} - m_0 c^2 + m_0 c^2\sqrt{1-\frac{v^2}{c^2}}$$

$$= \frac{m_0 v^2 + m_0 c^2 \left(1 - \dfrac{v^2}{c^2}\right)}{\sqrt{1 - \dfrac{v^2}{c^2}}} - m_0 c^2$$

$$K = \frac{m_0 c^2}{\sqrt{1 - \dfrac{v^2}{c^2}}} - m_0 c^2$$

When $v \langle\langle c, \dfrac{1}{\sqrt{1 - v^2/c^2}} = 1 + \dfrac{1}{2}\dfrac{v^2}{c^2}$ so, the Kinetic Energy $K = \dfrac{1}{2}m_0 v^2$ Thus the kinetic energy

approaches the classical limit for $v \langle\langle c$

By taking $m = \dfrac{m_0}{\sqrt{1 - \dfrac{v^2}{c^2}}}$, equation $K = \dfrac{m_0 c^2}{\sqrt{1 - \dfrac{v^2}{c^2}}} - m_0 c^2$ can be written as

$$K = mc^2 - m_0 c^2$$

$$\text{or } mc^2 = K + m_0 c^2$$

$$\text{or } E = K + m_0 c^2$$

Total energy = Kinetic energy + Rest mass energy.

Where $E = mc^2$ is the total energy. The first term in the right hand side arises due to external work done on the particle and the second term is due to rest mass energy.

The total energy equation can also be written in the form of linear momentum.

$$E = mc^2 = \frac{m_0 c^2}{\sqrt{1 - \dfrac{v^2}{c^2}}}, \quad E^2 = \frac{m_0^2 c^4}{\left(1 - \dfrac{v^2}{c^2}\right)}$$

$$\bar{p} = \frac{m_0 \bar{v}}{\sqrt{1 - \dfrac{v^2}{c^2}}}, \quad p^2 = \frac{m_0^2 v^2}{\left(1 - \dfrac{v^2}{c^2}\right)}$$

From above equations,

$$E^2 - p^2 c^2 = \frac{m_0^2 c^4}{\left(1 - v^2/c^2\right)} - \frac{m_0^2 v^2 c^2}{\left(1 - v^2/c^2\right)} = m_0^2 c^2 \frac{\left(c^2 - v^2\right)}{1 - v^2/c^2}$$

$$= m_0^2 c^4 \qquad\qquad [Taking\ v \langle\langle c]$$

$$or\ E^2 = p^2 c^2 + m_0^2 c^4$$

$$E = \sqrt{p^2 c^2 + m_0^2 c^4}$$

The above equation is another form of total energy. For massless particles such as neutrino, etc. the total energy $E = pc$

Reference

- Special_relativity_an_introduction: newworldencyclopedia.org, Retrieved 14 August, 2019

- VectorsTensors: physics.mq.edu.au, Retrieved 8 May, 2019

- Time-dilation, science: britannica.com, Retrieved 2 January, 2019

Quantum Mechanics

The theory of physics that describes nature at the smallest scales of energy levels of atoms and subatomic particles is known as quantum mechanics. It is a fundamental theory in physics that is applied in quantum computing, electronics, cryptography, quantum theory and macroscale quantum effects. The diverse applications of quantum mechanics in the current scenario have been thoroughly discussed in this chapter.

Quantum mechanics deals, with the behaviour of matter and light on the atomic and subatomic scale. It attempts to describe and account for the properties of molecules and atoms and their constituents—electrons, protons, neutrons, and other more esoteric particles such as quarks and gluons. These properties include the interactions of the particles with one another and with electromagnetic radiation (i.e., light, X-rays, and gamma rays).

The behaviour of matter and radiation on the atomic scale often seems peculiar, and the consequences of quantum theory are accordingly difficult to understand and to believe. Its concepts frequently conflict with common-sense notions derived from observations of the everyday world. There is no reason, however, why the behaviour of the atomic world should conform to that of the familiar, large-scale world. It is important to realize that quantum mechanics is a branch of physics and that the business of physics is to describe and account for the way the world—on both the large and the small scale—actually is and not how one imagines it or would like it to be.

The study of quantum mechanics is rewarding for several reasons. First, it illustrates the essential methodology of physics. Second, it has been enormously successful in giving correct results in practically every situation to which it has been applied. There is, however, an intriguing paradox. In spite of the overwhelming practical success of quantum mechanics, the foundations of the subject contain unresolved problems—in particular, problems concerning the nature of measurement. An essential feature of quantum mechanics is that it is generally impossible, even in principle, to measure a system without disturbing it; the detailed nature of this disturbance and the exact point at which it occurs are obscure and controversial.

Basic Concepts and Methods

Bohr's theory, which assumed that electrons moved in circular orbits, was extended by the German physicist Arnold Sommerfeld and others to include elliptic orbits and other refinements. Attempts were made to apply the theory to more complicated systems than the hydrogen atom. However, the ad hoc mixture of classical and quantum ideas made the theory and calculations increasingly unsatisfactory. Then, in the 12 months started in July 1925, a period of creativity without parallel in the history of physics, there appeared a series of papers by German scientists that set the subject on a firm conceptual foundation. The papers took two approaches: (1) matrix mechanics, proposed by Werner Heisenberg, Max Born, and Pascual Jordan, and (2) wave mechanics, put forward by Erwin Schrödinger. The protagonists were not always polite to each other. Heisenberg found the physical ideas of Schrödinger's theory "disgusting," and Schrödinger was "discouraged and re-pelled" by the lack of visualization in Heisenberg's method. However, Schrödinger, not allowing his emotions to interfere with his scientific endeavours, showed that, in spite of apparent dissimi-larities, the two theories are equivalent mathematically. The present discussion follows Schröding-er's wave mechanics because it is less abstract and easier to understand than Heisenberg's matrix mechanics.

Schrödinger's Wave Mechanics

Schrödinger expressed de Broglie's hypothesis concerning the wave behaviour of matter in a math-ematical form that is adaptable to a variety of physical problems without additional arbitrary as-sumptions. He was guided by a mathematical formulation of optics, in which the straight-line propagation of light rays can be derived from wave motion when the wavelength is small compared to the dimensions of the apparatus employed. In the same way, Schrödinger set out to find a wave equation for matter that would give particle-like propagation when the wavelength becomes com-paratively small. According to classical mechanics, if a particle of mass m_e is subjected to a force such that its potential energy is V(x, y, z) at position x, y, z, then the sum of V(x, y, z) and the kinetic energy $p^2/2m_e$ is equal to a constant, the total energy E of the particle. Thus,

$$\frac{P^2}{2m_e} + V(x, y, z) = E$$

It is assumed that the particle is bound—i.e., confined by the potential to a certain region in space because its energy E is insufficient for it to escape. Since the potential varies with position, two other quantities do also: the momentum and, hence, by extension from the de Broglie relation, the wavelength of the wave. Postulating a wave function $\Psi(x, y, z)$ that varies with position, Schröding-er replaced p in the above energy equation with a differential operator that embodied the de Brog-lie relation. He then showed that Ψ satisfies the partial differential equation

$$-\frac{\hbar^2}{2m_e}\left(\frac{\partial^2 \psi}{\partial x^2} + \frac{\partial^2 \psi}{\partial y^2} + \frac{\partial^2 \psi}{\partial z^2}\right) + V(x,y,z)\psi = E\psi$$

This is the (time-independent) Schrödinger wave equation, which established quantum mechan-ics in a widely applicable form. An important advantage of Schrödinger's theory is that no further arbitrary quantum conditions need be postulated. The required quantum results follow from cer-tain reasonable restrictions placed on the wave function—for example, that it should not become infinitely large at large distances from the centre of the potential.

Schrödinger applied his equation to the hydrogen atom, for which the potential function, given by classical electrostatics, is proportional to −e2/r, where −e is the charge on the electron. The nucleus (a proton of charge e) is situated at the origin, and r is the distance from the origin to the position of the electron. Schrödinger solved the equation for this particular potential with straight-forward, though not elementary, mathematics. Only certain discrete values of E lead to acceptable functions Ψ. These functions are characterized by a trio of integers n, l, m, termed quantum num-bers. The values of E depend only on the integers n (1, 2, 3, etc.) and are identical with those given by the Bohr theory. The quantum numbers l and m are related to the angular momentum of the electron; Square root of$\sqrt{l(l + 1)}\hbar$ is the magnitude of the angular momentum, and $m\hbar$ is its component along some physical direction.

The square of the wave function, Ψ^2, has a physical interpretation. Schrödinger originally supposed that the electron was spread out in space and that its density at point x, y, z was given by the value of Ψ^2 at that point. Almost immediately Born proposed what is now the accepted interpretation—namely, that Ψ^2 gives the probability of finding the electron at x, y, z. The distinction between the two interpretations is important. If Ψ^2 is small at a particular position, the original interpretation implies that a small fraction of an electron will always be detected there. In Born's interpretation, nothing will be detected there most of the time, but, when something is observed, it will be a whole electron. Thus, the concept of the electron as a point particle moving in a well-defined path around the nucleus is replaced in wave mechanics by clouds that describe the probable locations of electrons in different states.

Electron Spin and Antiparticles

In 1928 the English physicist Paul A.M. Dirac produced a wave equation for the electron that combined relativity with quantum mechanics. Schrödinger's wave equation does not satisfy the requirements of the special theory of relativity because it is based on a nonrelativistic expression for the kinetic energy ($p^2/2m_e$). Dirac showed that an electron has an additional quantum number m_s. Unlike the first three quantum numbers, m_s is not a whole integer and can have only the values $^{+1}/2$ and $^{-1}/2$. It corresponds to an additional form of angular momentum ascribed to a spinning motion. (The angular momentum mentioned above is due to the orbital motion of the electron, not its spin.) The concept of spin angular momentum was introduced in 1925 by Samuel A. Goudsmit and George E. Uhlenbeck, two graduate students at the University of Leiden, Neth., to explain the magnetic moment measurements made by Otto Stern and Walther Gerlach of Germany several years earlier. The magnetic moment of a particle is closely related to its angular momentum; if the angular momentum is zero, so is the magnetic moment. Yet Stern and Gerlach had observed a magnetic moment for electrons in silver atoms, which were known to have zero orbital angular momentum. Goudsmit and Uhlenbeck proposed that the observed magnetic moment was attributable to spin angular momentum.

The electron-spin hypothesis not only provided an explanation for the observed magnetic moment but also accounted for many other effects in atomic spectroscopy, including changes in spectral lines in the presence of a magnetic field (Zeeman effect), doublet lines in alkali spectra, and fine structure (close doublets and triplets) in the hydrogen spectrum.

The Dirac equation also predicted additional states of the electron that had not yet been observed. Experimental confirmation was provided in 1932 by the discovery of the positron by the American physicist Carl David Anderson. Every particle described by the Dirac equation has to have a corresponding antiparticle, which differs only in charge. The positron is just such an antiparticle of the negatively charged electron, having the same mass as the latter but a positive charge.

Identical Particles and Multielectron Atoms

Because electrons are identical to (i.e., indistinguishable from) each other, the wave function of an atom with more than one electron must satisfy special conditions. The problem of identical particles does not arise in classical physics, where the objects are large-scale and can always be distinguished, at least in principle. There is no way, however, to differentiate two electrons in the same atom, and the form of the wave function must reflect this fact. The overall wave function Ψ of a system of identical particles depends on the coordinates of all the particles. If the coordinates of two of the particles are interchanged, the wave function must remain unaltered or, at most, undergo a change of sign; the change of sign is permitted because it is Ψ^2 that occurs in the physical interpretation of the wave function. If the sign of Ψ remains unchanged, the wave function is said to be symmetric with respect to interchange; if the sign changes, the function is antisymmetric.

The symmetry of the wave function for identical particles is closely related to the spin of the particles. In quantum field theory, it can be shown that particles with half-integral spin ($1/2$, $3/2$, etc.) have antisymmetric wave functions. They are called fermions after the Italian-born physicist Enrico Fermi. Examples of fermions are electrons, protons, and neutrons, all of which have spin $1/2$. Particles with zero or integral spin (e.g., mesons, photons) have symmetric wave functions and are called bosons after the Indian mathematician and physicist Satyendra Nath Bose, who first applied the ideas of symmetry to photons in 1924–25.

The requirement of antisymmetric wave functions for fermions leads to a fundamental result, known as the exclusion principle, first proposed in 1925 by the Austrian physicist Wolfgang Pauli. The exclusion principle states that two fermions in the same system cannot be in the same quantum state. If they were, interchanging the two sets of coordinates would not change the wave function at all, which contradicts the result that the wave function must change sign. Thus, two electrons in the same atom cannot have an identical set of values for the four quantum numbers n, l, m, ms. The exclusion principle forms the basis of many properties of matter, including the periodic classification of the elements, the nature of chemical bonds, and the behaviour of electrons in solids; the last determines in turn whether a solid is a metal, an insulator, or a semiconductor.

The Schrödinger equation cannot be solved precisely for atoms with more than one electron. The principles of the calculation are well understood, but the problems are complicated by the number of particles and the variety of forces involved. The forces include the electrostatic forces between the nucleus and the electrons and between the electrons themselves, as well as weaker magnetic forces arising from the spin and orbital motions of the electrons. Despite these difficulties, approximation methods introduced by the English physicist Douglas R. Hartree, the Russian physicist Vladimir Fock, and others in the 1920s and 1930s have achieved considerable success. Such schemes start by assuming that each electron moves independently in an average electric field because of the nucleus and the other electrons; i.e., correlations between the positions of the electrons are ignored. Each electron has its own wave function, called an orbital. The overall wave function for all the electrons in the atom satisfies the exclusion principle. Corrections to the calculated energies are then made, which depend on the strengths of the electron-electron correlations and the magnetic forces.

Time-dependent Schrödinger Equation

At the same time that Schrödinger proposed his time-independent equation to describe the stationary states, he also proposed a time-dependent equation to describe how a system changes from one state to another. By replacing the energy E in Schrödinger's equation with a time-derivative operator, he generalized his wave equation to determine the time variation of the wave function as well as its spatial variation. The time-dependent Schrödinger equation reads

$$-\frac{\hbar^2}{2m_e}\left(\frac{\partial^2 \psi}{\partial x^2}+\frac{\partial^2 \psi}{\partial y^2}+\frac{\partial^2 \psi}{\partial z^2}\right)+V(x,y,z)\psi=i\hbar\frac{\partial \psi}{\partial t}.$$

The quantity i is the square root of −1. The function Ψ varies with time t as well as with position x, y, z. For a system with constant energy, E, Ψ has the form

$$\Psi(x,y,z,t)=\Psi(x,y,z)\exp\left(-\frac{iEt}{h}\right).$$

where exp stands for the exponential function, and the time-dependent Schrödinger equation reduces to the time-independent form.

The probability of a transition between one atomic stationary state and some other state can be calculated with the aid of the time-dependent Schrödinger equation. For example, an atom may change spontaneously from one state to another state with less energy, emitting the difference in energy as a photon with a frequency given by the Bohr relation. If electromagnetic radiation is applied to a set of atoms and if the frequency of the radiation matches the energy difference between two stationary states, transitions can be stimulated. In a stimulated transition, the energy of the atom may increase—i.e., the atom may absorb a photon from the radiation—or the energy of the atom may decrease, with the emission of a photon, which adds to the energy of the radiation. Such stimulated emission processes form the basic mechanism for the operation of lasers. The probability of a transition from one state to another depends on the values of the l, m, m_s quantum numbers of the initial and final states. For most values, the transition probability is effectively zero. However, for certain changes in the quantum numbers, summarized as selection rules, there is a finite probability. For example, according to one important selection rule, the l value changes by unity because photons have a spin of 1. The selection rules for radiation relate to the angular momentum properties of the stationary states. The absorbed or emitted photon has its own angular momentum, and the selection rules reflect the conservation of angular momentum between the atoms and the radiation.

Tunneling

The phenomenon of tunneling, which has no counterpart in classical physics, is an important consequence of quantum mechanics. Consider a particle with energy E in the inner region of a one-dimensional potential well V(x), (A potential well is a potential that has a lower value in a certain region of space than in the neighbouring regions.) In classical mechanics, if $E < V_0$ (the maximum height of the potential barrier), the particle remains in the well forever; if $E > V_0$, the particle escapes. In quantum mechanics, the situation is not so simple. The particle can escape even if its energy E is below the height of the barrier V_0, although the probability of escape is small unless E is close to V_0. In that case, the particle may tunnel through the potential barrier and emerge with the same energy E.

The phenomenon of tunneling has many important applications. For example, it describes a type of radioactive decay in which a nucleus emits an alpha particle (a helium nucleus). According to the quantum explanation given independently by George Gamow and by Ronald W. Gurney and Edward Condon in 1928, the alpha particle is confined before the decay by a potential of the shape. For a given nuclear species, it is possible to measure the energy E of the emitted alpha particle and the average lifetime τ of the nucleus before decay. The lifetime of the nucleus is a measure of the probability of tunneling through the barrier—the shorter the lifetime, the higher the probability. With plausible assumptions about the general form of the potential function, it is possible to calculate a relationship between τ and E that is applicable to all alpha emitters. This theory, which is borne out by experiment, shows that the probability of tunneling, and hence the value of τ, is extremely sensitive to the value of E. For all known alpha-particle emitters, the value of E varies from about 2 to 8 million electron volts, or MeV (1 MeV $= 10^6$ electron volts). Thus, the value of E varies only by a factor of 4, whereas the range of τ is from about 10^{11} years down to about $10-6$ second, a factor of 10^{24}. It would be difficult to account for this sensitivity of τ to the value of E by any theory other than quantum mechanical tunneling.

Axiomatic Approach

Although the two Schrödinger equations form an important part of quantum mechanics, it is possible to present the subject in a more general way. Dirac gave an elegant exposition of an axiomatic approach based on observables and states in a classic textbook entitled The Principles of Quantum Mechanics. An observable is anything that can be measured—energy, position, a component of angular momentum, and so forth. Every observable has a set of states, each state being represented by an algebraic function. With each state is associated a number that gives the result of a measurement of the observable. Consider an observable with N states, denoted by $\psi_1, \psi_2, \ldots, \psi_N$, and corresponding measurement values $a_1, a_2, \ldots, a_N$. A physical system—e.g., an atom in a particular state—is represented by a wave function Ψ, which can be expressed as a linear combination, or mixture, of the states of the observable. Thus, the Ψ may be written as

$$\Psi = c_1\Psi_1 + c_2\Psi_2 + \ldots + c_N\Psi_N.$$

For a given Ψ, the quantities c_1, c_2, etc., are a set of numbers that can be calculated. In general, the numbers are complex, but, in the present discussion, they are assumed to be real numbers.

The theory postulates, first, that the result of a measurement must be an a-value—i.e., a_1, a_2, or a_3, etc. No other value is possible. Second, before the measurement is made, the probability of obtaining the value a_1 is c_1^2, and that of obtaining the value a_2 is c_2^2, and so on. If the value obtained is, say, a_5, the theory asserts that after the measurement the state of the system is no longer the original Ψ but has changed to ψ_5, the state corresponding to a_5.

A number of consequences follow from these assertions. First, the result of a measurement cannot be predicted with certainty. Only the probability of a particular result can be predicted, even though the initial state (represented by the function Ψ) is known exactly. Second, identical measurements made on a large number of identical systems, all in the identical state Ψ, will produce different values for the measurements. This is, of course, quite contrary to classical physics and common sense, which say that the same measurement on the same object in the same state must produce the same result. Moreover, according to the theory, not only does the act of measurement change the state of the system, but it does so in an indeterminate way. Sometimes it changes the state to ψ_1, sometimes to ψ^2, and so forth.

There is an important exception to the above statements. Suppose that, before the measurement is made, the state Ψ happens to be one of the ψs—say, $\Psi = \psi_3$. Then $c_3 = 1$ and all the other cs are zero. This means that, before the measurement is made, the probability of obtaining the value a_3 is unity and the probability of obtaining any other value of a is zero. In other words, in this particular case, the result of the measurement can be predicted with certainty. Moreover, after the measurement is made, the state will be ψ_3, the same as it was before. Thus, in this particular case, measurement does not disturb the system. Whatever the initial state of the system, two measurements made in rapid succession (so that the change in the wave function given by the time-dependent Schrödinger equation is negligible) produce the same result.

The value of one observable can be determined by a single measurement. The value of two observables for a given system may be known at the same time, provided that the two observables have the same set of state functions $\psi_1, \psi_2, \ldots, \psi_N$. In this case, measuring the first observable results in a state function that is one of the ψs. Because this is also a state function of the second observable, the result of measuring the latter can be predicted with certainty. Thus the values of both observables are known. (Although the ψs are the same for the two observables, the two sets of a values are, in general, different.) The two observables can be measured repeatedly in any sequence. After the first measurement, none of the measurements disturbs the system, and a unique pair of values for the two observables is obtained.

Incompatible Observables

The measurement of two observables with different sets of state functions is a quite different situation. Measurement of one observable gives a certain result. The state function after the measurement is, as always, one of the states of that observable; however, it is not a state function for the second observable. Measuring the second observable disturbs the system, and the state of the system is no longer one of the states of the first observable. In general, measuring the first observable again does not produce the same result as the first time. To sum up, both quantities cannot be known at the same time, and the two observables are said to be incompatible.

A specific example of this behaviour is the measurement of the component of angular momentum along two mutually perpendicular directions. The Stern-Gerlach experiment mentioned above involved measuring the angular momentum of a silver atom in the ground state. In reconstructing this experiment, a beam of silver atoms is passed between the poles of a magnet. The poles are shaped so that the magnetic field varies greatly in strength over a very small distance (Figure). The apparatus determines the m_s quantum number, which can be +1/2 or −1/2. No other values are obtained. Thus in this case the observable has only two states—i.e., N = 2. The inhomogeneous magnetic field produces a force on the silver atoms in a direction that depends on the spin state of the atoms. The result is shown schematically in Figure. A beam of silver atoms is passed through magnet A. The atoms in the state with m_s = +1/2 are deflected upward and emerge as beam 1, while those with m_s = −1/2 are deflected downward and emerge as beam 2. If the direction of the magnetic field is the x-axis, the apparatus measures S_x, which is the x-component of spin angular momentum. The atoms in beam 1 have $S_x = +\hbar/2$ while those in beam 2 have $S_x = -\hbar/2$. In a classical picture, these two states represent atoms spinning about the direction of the x-axis with opposite senses of rotation.

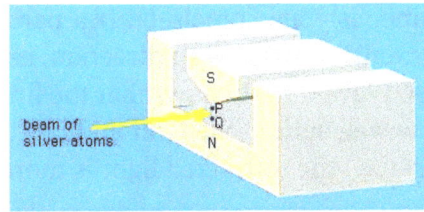

Figure: Magnet in Stern-Gerlach experiment. N and S are the north and south poles of a magnet.
The knife-edge of S results in a much stronger magnetic field at the point P than at Q.

The y-component of spin angular momentum S_y also can have only the values $+\hbar/2$ and $-\hbar/2$; however, the two states of S_y are not the same as for S_x. In fact, each of the states of S_x is an equal mixture of the states for S_y, and conversely. Again, the two S_y states may be pictured as representing atoms with opposite senses of rotation about the y-axis. These classical pictures of quantum states are helpful, but only up to a certain point. For example, quantum theory says that each of the states corresponding to spin about the x-axis is a superposition of the two states with spin about the y-axis. There is no way to visualize this; it has absolutely no classical counterpart. One simply has to accept the result as a consequence of the axioms of the theory. Suppose that, as in Figure, the atoms in beam 1 are passed into a second magnet B, which has a magnetic field along the y-axis perpendicular to x. The atoms emerge from B and go in equal numbers through its two output channels. Classical theory says that the two magnets together have measured both the x- and y-components of spin angular momentum and that the atoms in beam 3 have $S_x = +\hbar/2$, $S_y = +\hbar/2$, while those in beam 4 have $S_x = +\hbar/2$, $S_y = -\hbar/2$. However, classical theory is wrong, because if beam 3 is put through still another magnet C, with its magnetic field along x, the atoms divide equally into beams 5 and 6 instead of emerging as a single beam 5 (as they would if they had $S_x = +\hbar/2$). Thus, the correct statement is that the beam entering B has $S_x = +\hbar/2$ and is composed of an equal mixture of the states $S_y = +\hbar/2$ and $S_y = -\hbar/2$—i.e., the x-component of angular momentum is known but the y-component is not. Correspondingly, beam 3 leaving B has $S_y = +\hbar/2$ and is an equal mixture of the states $S_x = +\hbar/2$ and $S_x = -\hbar/2$; the y-component of angular momentum is known but the x-component is not. The information about S_x is lost because of the disturbance caused by magnet B in the measurement of S_y.

Heisenberg Uncertainty Principle

The observables discussed so far have had discrete sets of experimental values. For example, the values of the energy of a bound system are always discrete, and angular momentum components have values that take the form $m\hbar$, where m is either an integer or a half-integer, positive or negative. On the other hand, the position of a particle or the linear momentum of a free particle can take continuous values in both quantum and classical theory. The mathematics of observables with a continuous spectrum of measured values is somewhat more complicated than for the discrete case but presents no problems of principle. An observable with a continuous spectrum of measured values has an infinite number of state functions. The state function Ψ of the system is still regarded as a combination of the state functions of the observable, but the sum in equation $\Psi = c_1\Psi_1 + c_2\Psi_2 + \ldots + c_N\Psi_N$. must be replaced by an integral.

Measurements can be made of position x of a particle and the x-component of its linear momentum, denoted by p_x. These two observables are incompatible because they have different state functions. The phenomenon of diffraction noted above illustrates the impossibility of measuring position and momentum simultaneously and precisely. If a parallel monochromatic light beam passes through a slit Figure, its intensity varies with direction, as shown in Figure. The light has zero intensity in certain directions. Wave theory shows that the first zero occurs at an angle θ_o, given by $\sin \theta_o = \lambda/b$, where λ is the wavelength of the light and b is the width of the slit. If the width of the slit is reduced, θ_o increases—i.e., the diffracted light is more spread out. Thus, θ_o measures the spread of the beam.

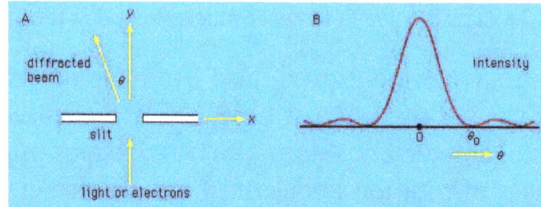

Figure: Parallel monochromatic light incident normally on a slit, (B) variation in the intensity of the light with direction after it has passed through the slit. If the experiment is repeated with electrons instead of light, the same diagram would represent the variation in the intensity (i.e., relative number) of the electrons.

The experiment can be repeated with a stream of electrons instead of a beam of light. According to de Broglie, electrons have wavelike properties; therefore, the beam of electrons emerging from the slit should widen and spread out like a beam of light waves. This has been observed in experiments. If the electrons have velocity u in the forward direction (i.e., the y-direction in Figure), their (linear) momentum is $p = m_e u$. Consider p_x, the component of momentum in the x-direction. After the electrons have passed through the aperture, the spread in their directions results in an uncertainty in p_x by an amount

$$\Delta p_x \approx p \, \sin \theta_0 = p \frac{\lambda}{b}$$

where λ is the wavelength of the electrons and, according to the de Broglie formula, equals h/p. Thus, $\Delta p_x \approx h/b$. Exactly where an electron passed through the slit is unknown; it is only certain that an electron went through somewhere. Therefore, immediately after an electron goes through, the uncertainty in its x-position is $\Delta x \approx b/2$. Thus, the product of the uncertainties is of the order of $\hbar$. More exact analysis shows that the product has a lower limit, given by

$$\Delta x \Delta px \geq \frac{h}{2}.$$

This is the well-known Heisenberg uncertainty principle for position and momentum. It states that there is a limit to the precision with which the position and the momentum of an object can be measured at the same time. Depending on the experimental conditions, either quantity can be measured as precisely as desired (at least in principle), but the more precisely one of the quantities is measured, the less precisely the other is known.

The uncertainty principle is significant only on the atomic scale because of the small value of h in everyday units. If the position of a macroscopic object with a mass of, say, one gram is measured with a precision of 10^{-6} metre, the uncertainty principle states that its velocity cannot be measured to better than about 10^{-25} metre per second. Such a limitation is hardly worrisome. However, if an electron is located in an atom about 10^{-10} metre across, the principle gives a minimum uncertainty in the velocity of about 10^6 metre per second.

The above reasoning leading to the uncertainty principle is based on the wave-particle duality of the electron. When Heisenberg first propounded the principle in 1927 his reasoning was based, however, on the wave-particle duality of the photon. He considered the process of measuring the position of an electron by observing it in a microscope. Diffraction effects due to the wave nature of light result in a blurring of the image; the resulting uncertainty in the position of the electron is approximately equal to the wavelength of the light. To reduce this uncertainty, it is necessary to use light of shorter wavelength—e.g., gamma rays. However, in producing an image of the electron, the gamma-ray photon bounces off the electron, giving the Compton effect. As a result of the colli-sion, the electron recoils in a statistically random way. The resulting uncertainty in the momentum of the electron is proportional to the momentum of the photon, which is inversely proportional to the wavelength of the photon. So it is again the case that increased precision in knowledge of the position of the electron is gained only at the expense of decreased precision in knowledge of its momentum. Heisenberg's reasoning brings out clearly the fact that the smaller the particle being observed, the more significant is the uncertainty principle. When a large body is observed, photons still bounce off it and change its momentum, but, considered as a fraction of the initial momentum of the body, the change is insignificant.

The Schrödinger and Dirac theories give a precise value for the energy of each stationary state, but in reality the states do not have a precise energy. The only exception is in the ground (lowest energy) state. Instead, the energies of the states are spread over a small range. The spread arises from the fact that, because the electron can make a transition to another state, the initial state has a finite lifetime. The transition is a random process, and so different atoms in the same state have different lifetimes. If the mean lifetime is denoted as τ, the theory shows that the energy of the initial state has a spread of energy ΔE, given by

$$\tau \Delta E \approx h$$

This energy spread is manifested in a spread in the frequencies of emitted radiation. Therefore, the spectral lines are not infinitely sharp. (Some experimental factors can also broaden a line, but their effects can be reduced; however, the present effect, known as natural broadening, is fundamental and cannot be reduced.) Equation is another type of Heisenberg uncertainty relation; generally, if a measurement with duration τ is made of the energy in a system, the measurement disturbs the system, causing the energy to be uncertain by an amount ΔE, the magnitude of which is given by the above equation.

Quantum Entanglement

Quantum entanglement is one of the central principles of quantum physics, though it is also highly misunderstood. In short, quantum entanglement means that multiple particles are linked together in a way such that the measurement of one particle's quantum state determines the possible quantum states of the other particles. This connection isn't depending on the location of the particles in space. Even if you separate entangled particles by billions of miles, changing one particle will induce a change in the other. Even though quantum entanglement appears to transmit information instantaneously, it doesn't actually violate the classical speed of light because there's no "movement" through space.

The Classic Quantum Entanglement Example

The classic example of quantum entanglement is called the EPR paradox. In a simplified version of this case, consider a particle with quantum spin 0 that decays into two new particles, Particle A and Particle B. Particle A and Particle B head off in opposite directions. However, the original particle had a quantum spin of 0. Each of the new particles has a quantum spin of 1/2, but because they have to add up to 0, one is +1/2 and one is -1/2.

This relationship means that the two particles are entangled. When you measure the spin of Particle A, that measurement has an impact on the possible results you could get when measuring the spin of Particle B. And this isn't just an interesting theoretical prediction but has been verified experimentally through tests of Bell's Theorem.

One important thing to remember is that in quantum physics, the original uncertainty about the particle's quantum state isn't just a lack of knowledge. A fundamental property of quantum theory is that prior to the act of measurement, the particle really doesn't have a definite state, but is in a superposition of all possible states. This is best modeled by the classic quantum physics thought experiment, Schroedinger's Cat, where a quantum mechanics approach results in an unobserved cat that is both alive and dead simultaneously.

Quantum Field Theory

Quantum field theory is the body of physical principles combining the elements of quantum mechanics with those of relativity to explain the behaviour of subatomic particles and their interactions via a variety of force fields. Two examples of modern quantum field theories are quantum electrodynamics, describing the interaction of electrically charged particles and the electromagnetic force, and quantum chromodynamics, representing the interactions of quarks and the strong force. Designed to account for particle-physics phenomena such as high-energy collisions in which subatomic particles may be created or destroyed, quantum field theories have also found applications in other branches of physics.

The prototype of quantum field theories is quantum electrodynamics (QED), which provides a comprehensive mathematical framework for predicting and understanding the effects of electromagnetism on electrically charged matter at all energy levels. Electric and magnetic forces are regarded as arising from the emission and absorption of exchange particles called photons. These can be represented as disturbances of electromagnetic fields, much as ripples on a lake are disturbances of the water. Under suitable conditions, photons may become entirely free of charged particles; they are then detectable as light and as other forms of electromagnetic radiation. Similarly, particles such as electrons are themselves regarded as disturbances of their own quantized fields. Numerical predictions based on QED agree with experimental data to within one part in 10 million in some cases.

There is a widespread conviction among physicists that other forces in nature—the weak force responsible for radioactive beta decay; the strong force, which binds together the constituents of atomic nuclei; and perhaps also the gravitational force—can be described by theories similar to QED. These theories are known collectively as gauge theories. Each of the forces is mediated by its own set of exchange particles, and differences between the forces are reflected in the properties of these particles. For example, electromagnetic and gravitational forces operate over long distances, and their exchange particles—the well-studied photon and the as-yet-undetected graviton, respectively—have no mass.

In contrast, the strong and weak forces operate only over distances shorter than the size of an atomic nucleus. Quantum chromodynamics (QCD), the modern quantum field theory describing the effects of the strong force among quarks, predicts the existence of exchange particles called gluons, which are also massless as with QED but whose interactions occur in a way that essentially confines quarks to bound particles such as the proton and the neutron. The weak force is carried by massive exchange particles—the W and Z particles—and is thus limited to an extremely short range, approximately 1 percent of the diameter of a typical atomic nucleus.

The current theoretical understanding of the fundamental interactions of matter is based on quantum field theories of these forces. Research continues, however, to develop a single unified field theory that encompasses all the forces. In such a unified theory, all the forces would have a common origin and would be related by mathematical symmetries. The simplest result would be that all the forces would have identical properties and that a mechanism called spontaneous symmetry breaking would account for the observed differences. A unified theory of electromagnetic and weak forces, the electroweak theory, has been developed and has received considerable experimental support. It is likely that this theory can be extended to include the strong force. There also exist theories that include the gravitational force, but these are more speculative.

Quantum Electrodynamics

Quantum electrodynamics, commonly referred to as QED, is a quantum field theory of the electromagnetic force. Taking the example of the force between two electrons, the classical theory of electromagnetism would describe it as arising from the electric field produced by each electron at the position of the other. The force can be calculated from Coulomb's law.

The quantum field theory approach visualizes the force between the electrons as an exchange force arising from the exchange of virtual photons. It is represented by a series of Feynman diagrams, the most basic of which is

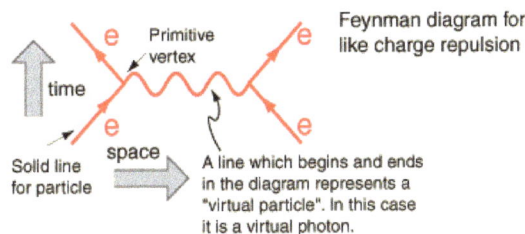

With time proceeding upward in the diagram, this diagram describes the electron interaction in which two electrons enter, exchange a photon, and then emerge. Using a mathematical approach

known as the Feynman calculus, the strength of the force can be calculated in a series of steps which assign contributions to each of the types of Feynman diagrams associated with the force.

QED applies to all electromagnetic phenomena associated with charged fundamental particles such as electrons and positrons, and the associated phenomena such as pair production, electron-positron annihilation, Compton scattering, etc. It was used to precisely model some quantum phenomena which had no classical analogs, such as the Lamb shift and the anomalous magnetic moment of the electron. QED was the first successful quantum field theory, incorporating such ideas as particle creation and annihilation into a self-consistent framework. The development of the theory was the basis of the 1965 Nobel Prize in physics, awarded to Richard Feynman, Julian Schwinger and Sin-itero Tomonaga.

Reference

- Quantum-mechanics-physics, science: britannica.com, Retrieved 11 February, 2019

- What-is-quantum-entanglement: thoughtco.com, Retrieved 1 July, 2019

- Quantum-field-theory, science: britannica.com, Retrieved 20 January, 2019

- Forces: phy-astr.gsu.edu, Retrieved 25 April, 2019

Atomic Physics

Atom is the smallest fundamental unit of ordinary matter with the properties of a chemical element. Atomic physics is the branch of physics that deals with the study of atoms as an isolated system of electrons and atomic nucleus. This chapter explains all the diverse aspects of atomic physics in order to develop a better understanding of the subject matter.

Atomic physics is the study of the structure of the atom, its energy states, and its interactions with other particles and with electric and magnetic fields. Atomic physics has proved to be a spectacularly successful application of quantum mechanics, which is one of the cornerstones of modern physics.

The notion that matter is made of fundamental building blocks dates to the ancient Greeks, who speculated that earth, air, fire, and water might form the basic elements from which the physical world is constructed. They also developed various schools of thought about the ultimate nature of matter. Perhaps the most remarkable was the atomist school founded by the ancient Greeks Leucippus of Miletus and Democritus of Thrace. They developed the notion that matter consists of indivisible and indestructible atoms. The atoms are in ceaseless motion through the surrounding void and collide with one another like billiard balls, much like the modern kinetic theory of gases. However, the necessity for a void (or vacuum) between the atoms raised new questions that could not be easily answered. For this reason, the atomist picture was rejected by Aristotle and the Athenian school in favour of the notion that matter is continuous. The idea nevertheless persisted, and it reappeared in the writings of the Roman poet Lucretius, in his work De rerum natura (On the Nature of Things).

Little more was done to advance the idea that matter might be made of tiny particles. The English physicist Isaac Newton, in his Principia Mathematica (1687), proposed that Boyle's law, which states that the product of the pressure and the volume of a gas is constant at the same temperature, could be explained if one assumes that the gas is composed of particles. English chemist John Dalton suggested that each element consists of identical atoms, and in 1811 the Italian physicist Amedeo Avogadro hypothesized that the particles of elements may consist of two or more atoms stuck together. Avogadro called such conglomerations molecules, and, on the basis of experimental work, he conjectured that the molecules in a gas of hydrogen or oxygen are formed from pairs of atoms.

During the 19th century there developed the idea of a limited number of elements, each consisting of a particular type of atom, that could combine in an almost limitless number of ways to form chemical compounds. At mid-century the kinetic theory of gases successfully attributed such phenomena as the pressure and viscosity of a gas to the motions of atomic and molecular particles. the growing weight of chemical evidence and the success of the kinetic theory left little doubt that atoms and molecules were real.

The internal structure of the atom, however, became clear only in the early 20th century with the work of the British physicist Ernest Rutherford and his students. Until Rutherford's efforts, a popular model of the atom had been the so-called "plum-pudding" model, advocated by the English physicist

Joseph John Thomson, which held that each atom consists of a number of electrons (plums) embedded in a gel of positive charge (pudding); the total negative charge of the electrons exactly balances the total positive charge, yielding an atom that is electrically neutral. Rutherford conducted a series of scattering experiments that challenged Thomson's model. Rutherford observed that when a beam of alpha particles (which are now known to be helium nuclei) struck a thin gold foil, some of the particles were deflected backward. Such large deflections were inconsistent with the plum-pudding model.

This work led to Rutherford's atomic model, in which a heavy nucleus of positive charge is surrounded by a cloud of light electrons. The nucleus is composed of positively charged protons and electrically neutral neutrons, each of which is approximately 1,836 times as massive as the electron. Because atoms are so minute, their properties must be inferred by indirect experimental techniques. Chief among these is spectroscopy, which is used to measure and interpret the electromagnetic radiation emitted or absorbed by atoms as they undergo transitions from one energy state to another. Each chemical element radiates energy at distinctive wavelengths, which reflect their atomic structure. Through the procedures of wave mechanics, the energies of atoms in various energy states and the characteristic wavelengths they emit may be computed from certain fundamental physical constants—namely, electron mass and charge, the speed of light, and Planck's constant. Based on these fundamental constants, the numerical predictions of quantum mechanics can account for most of the observed properties of different atoms. In particular, quantum mechanics offers a deep understanding of the arrangement of elements in the periodic table, showing, for example, that elements in the same column of the table should have similar properties.

In recent years the power and precision of lasers have revolutionized the field of atomic physics. On the one hand, lasers have dramatically increased the precision with which the characteristic wavelengths of atoms can be measured. For example, modern standards of time and frequency are based on measurements of transition frequencies in atomic cesium, and the definition of the metre as a unit of length is now related to frequency measurements through the velocity of light. In addition, lasers have made possible entirely new technologies for isolating individual atoms in electromagnetic traps and cooling them to near absolute zero. When the atoms are brought essentially to rest in the trap, they can undergo a quantum mechanical phase transition to form a superfluid known as a Bose-Einstein condensation, while remaining in the form of a dilute gas. In this new state of matter, all the atoms are in the same coherent quantum state. As a consequence, the atoms lose their individual identities, and their quantum mechanical wavelike properties become dominant. The entire condensate then responds to external influences as a single coherent entity (like a school of fish), instead of as a collection of individual atoms. Recent work has shown that a coherent beam of atoms can be extracted from the trap to form an "atom laser" analogous to the coherent beam of photons in a conventional laser. The atom laser is still in an early stage of development, but it has the potential to become a key element of future technologies for the fabrication of microelectronic and other nanoscale devices.

Atom

Atoms are the basic units of matter and the defining structure of elements. We now know that atoms are made up of three particles: protons, neutrons and electrons — which are composed of even smaller particles such as quarks.

Atoms were created after the Big Bang 13.7 billion years ago. As the hot, dense new universe cooled, conditions became suitable for quarks and electrons to form. Quarks came together to form protons and neutrons, and these particles combined into nuclei. This all took place within the first few minutes of the universe's existence.

It took 380,000 years for the universe to cool down enough to slow down the electrons so that the nuclei could capture them to form the first atoms. The earliest atoms were primarily hydrogen and helium, which are still the most abundant elements in the universe. Gravity eventually caused clouds of gas to coalesce and form stars, and heavier atoms were (and still are) created within the stars and sent throughout the universe when the star exploded (supernova).

Atomic Particles

Protons and neutrons are heavier than electrons and reside in the nucleus at the center of the atom. Electrons are extremely lightweight and exist in a cloud orbiting the nucleus. The electron cloud has a radius 10,000 times greater than the nucleus.

Protons and neutrons have approximately the same mass. However, one proton weighs more than 1,800 electrons. Atoms always have an equal number of protons and electrons, and the number of protons and neutrons is usually the same as well. Adding a proton to an atom makes a new element, while adding a neutron makes an isotope, or heavier version, of that atom.

Nucleus

The nucleus was discovered in 1911 by Ernest Rutherford, a physicist from New Zealand, who in 1920 proposed the name proton for the positively charged particles of the atom. Rutherford also theorized that there was also a neutral particle within the nucleus, which James Chadwick, a British physicist and student of Rutherford, was able to confirm in 1932.

Virtually all the mass of the atom resides in the nucleus. The protons and neutrons that make up the nucleus are approximately the same mass (the proton is slightly less) and have the same angular momentum.

The nucleus is held together by the "strong force," one of the four basic forces in nature. This force between the protons and neutrons overcomes the repulsive electrical force that would, according to the rules of electricity, push the protons apart otherwise. Some atomic nuclei are unstable because the binding force varies for different atoms based on the size of the nucleus. These atoms will then decay into other elements, such as carbon-14 decaying into nitrogen-14.

Protons

Protons are positively charged particles found within atomic nuclei. Rutherford discovered them in experiments with cathode-ray tubes conducted between 1911 and 1919. Protons are slightly smaller in mass than neutrons with a relative mass of 0.9986 (as compared with the mass of the neutron being 1) or about 1.673×10^{-27} kg.

The number of protons in an atom defines what element it is. For example, carbon atoms have six protons, hydrogen atoms have one and oxygen atoms have eight. The number of protons in an atom is referred to as the atomic number of that element. The number of protons in an atom also determines the chemical behavior of the element. The Periodic Table of the Elements arranges elements in order of increasing atomic number.

Three quarks make up each proton — two "up" quarks (each with a 2/3 positive charge) and one "down" quark (with a 1/3 negative charge) — and they are held together by other subatomic particles called gluons, which are massless.

Electrons

Electrons are tiny compared to protons and neutrons, over 1,800 times smaller than either a proton or a neutron. Electrons have a relative mass of 0.0005439 (as compared with the mass of a neutron being 1) or about 9.109×10^{-31} kg.

J.J. Thomson, a British physicist, discovered the electron in 1897. Originally known as "corpuscles," electrons have a negative charge and are electrically attracted to the positively charged protons. Electrons surround the atomic nucleus in pathways called orbitals, an idea that was put forth by Erwin Schrödinger, an Austrian physicist, in the 1920s. Today, this model is known as the quantum model or the electron cloud model. The inner orbitals surrounding the atom are spherical but the outer orbitals are much more complicated.

An atom's electron configuration is the orbital description of the locations of the electrons in a typical atom. Using the electron configuration and principles of physics, chemists can predict an atom's properties, such as stability, boiling point and conductivity.

Typically, only the outermost electron shells matter in chemistry. The inner electron shell notation is often truncated by replacing the longhand orbital description with the symbol for a noble gas in brackets. This method of notation vastly simplifies the description for large molecules.

For example, the electron configuration for beryllium (Be) is $1s^2 2s^2$, but it's is written $[He]2s^2$. [He] is equivalent to all the electron orbitals in a helium atom. The letters, s, p, d, and f designate the shape of the orbitals and the superscript gives the number of electrons in that orbital. Uranium, as another example, has an electron configuration of $1s^2 2s^2 2p^6 3s^2 3p^6 4s^2 3d^{10} 4p^6 5s^2 4d^{10} 5p^6 6s^2 4e^{14} 5d^{10} 6p^6 7s^2 5f^4$, which can be simplified to $[RN]7s^2 5f^4$.

Neutrons

The neutron is used as a comparison to find the relative mass of protons and electrons (so it has a relative mass of 1) and has a physical mass of 1.6749×10^{-27} kg.

The neutron's existence was theorized by Rutherford in 1920 and discovered by Chadwick in 1932. Neutrons were found during experiments when atoms were shot at a thin sheet of beryllium. Subatomic particles with no charge were released – the neutron.

Neutrons are uncharged particles found within all atomic nuclei (except for hydrogen-1). A neutron's mass is slightly larger than that of a proton. Like protons, neutrons are also made of quarks — one "up" quark (with a positive 2/3 charge) and two "down" quarks (each with a negative 1/3 charge).

Models of Atom

Soon after the discovery of the sub-atomic particles of an atom, scientists were eager to figure out the distribution of these particles within the atom. Several atomic models were proposed to explain the structure of the atom. However, a lot of them could not explain the stability of the atom.

Thomson's Atomic Model

In 1898, J. J. Thomson proposed the first of many atomic models to come. He proposed that an atom is shaped like a sphere with a radius of approximately 10^{-10}m, where the positive charge is uniformly distributed. The electrons are embedded in this sphere so as to give the most stable electrostatic arrangement.

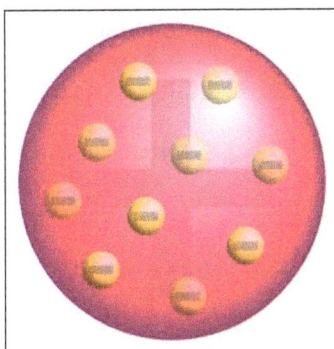

Thomsons' atomic model

Doesn't the figure above remind you of a cut watermelon with seeds inside? Or, you can also think of it as a pudding with the electrons being the plum or the raisins in the pudding. Therefore, this model is also referred to as the watermelon model, the plum pudding model or the raisin pudding model.

An important aspect of this model is that it assumes that the mass of the atom is uniformly distributed over the atom. Thomson's atomic model was successful in explaining the overall neutrality of the atom. However, its propositions were not consistent with the results of later experiments. In 1906, J. J. Thomson was awarded the Nobel Prize in physics for his theories and experiments on electricity conduction by gases.

Rutherford's Atomic Model

The second of the atomic models was the contribution of Ernest Rutherford. To come up with their model, Rutherford and his students – Hans Geiger and Ernest Marsden performed an experiment where they bombarded very thin gold foil with α-particles.

α-Particle Scattering Experiment

Experiment: In this experiment, high energy α-particles from a radioactive source were directed at a thin foil (about 100nm thickness) of gold. A circular, fluorescent zinc sulfide screen was present around the thin gold foil. A tiny flash of light was produced at a point on the screen whenever α-particles struck it.

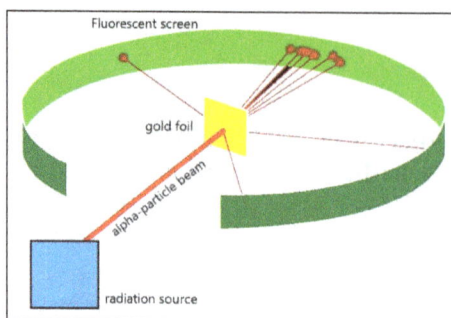

Rutherford's alpha-particle scattering experiment

Results: Based on Thomson's model, the mass of every atom in the gold foil should be evenly spread over the entire atom. Therefore, when α-particles hit the foil, it is expected that they would slow down and change directions only by small angles as they pass through the foil. However, the results from Rutherford's experiment were unexpected –

- Most of the α-particles passed undeflected through the foil.

- A small number of α-particles were deflected by small angles.

- Very few α-particles (about 1 in 20,000) bounced back.

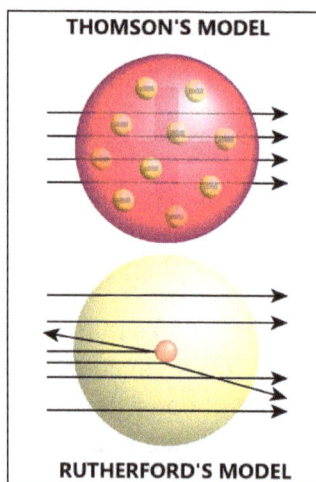

Thomson's model versus Rutherford's model

Based on the above results, Rutherford made the following conclusions about the structure of the atom:

- Since most of the α-particles passed through the foil undeflected, most of the space in the atom is empty.

- The deflection of a few positively charged α-particles must be due to the enormous repulsive force. This suggests that the positive charge is not uniformly spread throughout the atom as Thomson had proposed. The positive charge has to be concentrated in a very small volume to deflect the positively charged α-particles.

- Rutherford's calculations show that the volume of the nucleus is very small compared to the total volume of the atom. The radius of an atom is about 10-10m, while that of the nucleus is 10-15m.

Nuclear Model of the Atom

Based on his observations and conclusions, Rutherford proposed his model of the structure of the atom. According to this model –

- Most of the mass of the atom and the positive charge is densely concentrated in a very small region in the atom. Rutherford called this region the nucleus.

- Electrons surround the nucleus and move around it at very high speeds in circular paths called orbits. This arrangement also resembles the solar system, where the nucleus forms the sun and the electrons are the revolving planets. Therefore, it is also referred to as the Planetary Model.

- Electrostatic forces of attraction hold the nucleus and electrons together.

Drawbacks of Rutherford's Atomic Model

- According to Rutherford's atomic model, the electrons (planets) move around the nucleus (sun) in well-defined orbits. Since a body that moves in an orbit must undergo acceleration, the electrons, in this case, must be under acceleration. According to Maxwell's electromagnetic theory, charged particles when accelerated must emit electromagnetic radiation. Therefore, an electron in an orbit will emit radiation and eventually the orbit will shrink. If this is true, then the electron will spiral into the nucleus. But this does not happen. Thus, Rutherford's model does not explain the stability of the atom.

- Contrarily, let's consider that the electrons do not move and are stationary. Then the electrostatic attraction between the electrons and the dense nucleus will pull the electrons into the nucleus to form a miniature version of Thomson's model.

- Rutherford's model also does not state anything about the distribution of the electrons around the nucleus and the energies of these electrons.

Thus, Thomson and Rutherford's atomic models revealed key aspects of the structure of the atom but failed to address some critical points. Now that we know the two atomic models, let's try to understand a few concepts.

Atomic Number and Mass Number

As we know now, a positive charge on the nucleus is due to the protons. Also, the charge on the proton is equal but opposite to that of the electron. Atomic Number (Z) is the number of protons present in the nucleus. For example, the number of protons in sodium is 11 whereas it is 1 in hydrogen, Therefore, the atomic numbers of sodium and hydrogen are 11 and 1, respectively.

Also, to maintain electrical neutrality, the number of electrons in an atom is equal to the number of protons (atomic number, Z). Therefore, the number of electrons in sodium and hydrogen is 11 and 1, respectively.

Atomic number = the number of protons in the nucleus of an atom

= the number of electrons in a neutral atom

The positive charge on the nucleus is due to protons, but the mass of the atom is due to protons and neutrons. They are collectively known as nucleons. Mass number (A) of the atom is the total number of nucleons.

Mass number (A) = the number of protons (Z) + the number of neutrons (n)

Therefore, the composition of an atom is represented using the element symbol (X) with the mass number (A) as super-script on the left and atomic number (Z) as sub-script on the left $^{-A}_{Z}X$.

Isobars and Isotopes

Isobars are atoms with the same mass number but a different atomic number. For example, 146C and 147N.

Isotopes, on the other hand, are atoms with the same atomic number but a different mass number. This means that the difference in the isotopes is due to the presence of a different number of neutrons in the nucleus. Let's understand this using hydrogen as an example –

- 99.985% of hydrogen atoms contain only one proton. This isotope is protium ($^{1}_{1}H$).

- The isotope containing one proton and one neutron is deuterium ($^{2}_{1}D$).

- The isotope with one proton and two neutrons is tritium ($^{3}_{1}T$). This isotope exists in trace amounts on earth.

Other common isotopes are – carbon atoms with 6 protons and 6, 7, or 8 neutrons ($^{12}_{6}C, ^{13}_{6}C, ^{14}_{6}C$) and chlorine atoms with 17 protons and 18 or 20 neutrons ($^{35}_{17}Cl, ^{37}_{17}Cl$).

Note: Chemical properties of atoms are under the influence of the number of electrons, which are dependent on the number of protons in the nucleus. The number of neutrons has a very little effect on the chemical properties of an element. Therefore, all isotopes of an element show same chemical behaviour.

Bohr Model of Atom

The Bohr Model has an atom consisting of a small, positively-charged nucleus orbited by negatively-charged electrons. Here's a closer look at the Bohr Model, which is sometimes called the Rutherford-Bohr Model.

Niels Bohr proposed the Bohr Model of the Atom in 1915. Because the Bohr Model is a modification of the earlier Rutherford Model, some people call Bohr's Model the Rutherford-Bohr Model. The modern model of the atom is based on quantum mechanics. The Bohr Model contains some errors, but it is important because it describes most of the accepted features of atomic theory without all of the high-level math of the modern version. Unlike earlier models, the Bohr Model explains the Rydberg formula for the spectral emission lines of atomic hydrogen.

The Bohr Model is a planetary model in which the negatively-charged electrons orbit a small, positively-charged nucleus similar to the planets orbiting the Sun (except that the orbits are not planar). The gravitational force of the solar system is mathematically akin to the Coulomb (electrical) force between the positively-charged nucleus and the negatively-charged electrons.

Fundamental Postulates

The Danish physicist Niels Bohr, who first presented this model of the atom, based it on 3 fundamental postulates.

1. Electrons move around the nucleus in circular non-radiating orbits - called "stationary states". However, they are not at rest!

2. An atom only emits or absorbs electromagnetic radiation when an electron makes a transition from one state to another.

3. Only certain stationary states are allowed: those where the orbital angular momentum of the electron is given by $L = n = n\hbar / 2\pi$

where n is an integer ≥ 1 (n = 1, 2, 3 ... etc.) and h is Planck's constant.

This is known as the quantization of angular momentum.

The equation implies that an integer number of wavelengths fit round the orbit:

$$L = pr = nh / 2\pi \text{ whence } 2\pi r = nh / p = n\lambda$$

since, $\lambda = h / p$ (de Broglie relation)

Condition: Only waves with an integral number of de Broglie wavelengths around the orbit are allowed.

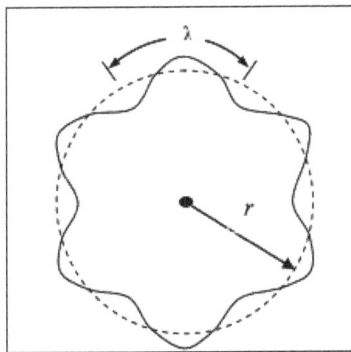

Given this condition, amplitudes of successive orbits will be in phase - reinforce each other and generate a standing-wave pattern. These are the allowed waves. The superposition of wave amplitudes of successive orbits with any other wavelength will average to zero and are not allowed.

A full quantum-mechanical calculation gives the same allowed states as the Bohr model. The general result then is that angular momentum is quantized, i.e. an integral number of a basic unit $h / 2\pi$.

Sizes of Allowed Orbits

Classically, for an orbiting electron (mass m, charge -e and speed v), the centripetal force is balanced by the electric (Coulomb) force:

$$\frac{mv^2}{r} = \frac{Ze^2}{4\pi\varepsilon_0 r^2}$$

For hydrogen, $Z = 1$, so :
$$\frac{mv^2}{r} - \frac{e^2}{4\pi\varepsilon_0 r^2}$$

From equation $L = n = n\hbar / 2\pi$:

$$\frac{mv^2}{r} = \frac{m^2 v^2 r^2}{mr^3} = \frac{(pr)^2}{mr^3} = \frac{L^2}{mr^3} = \frac{n^2 h^2}{4\pi^2 mr^3}$$

Substituting into equation $\frac{mv^2}{r} - \frac{e^2}{4\pi\varepsilon_0 r^2}$:

$$\frac{n^2 h^2}{4\pi^2 mr^3} - \frac{e^2}{4\pi\varepsilon_0 r^2}$$

whence, the orbit radius $r = n^2 \left(\frac{\varepsilon_0 h^2}{\pi m e^2} \right) = n^2 a_0$

a_0 is called the Bohr radius - radius of the innermost (n = 1) orbit of the hydrogen atom.

$$a_0 = \varepsilon_0 h^2 / \left(\pi m e^2 \right) = 5.29 \times 10^{-11} \, m = 0.529 \, \mathring{A}.$$

Energies of Allowed Orbits

First, consider hydrogen: Using equation $\frac{mv^2}{r} - \frac{e^2}{4\pi\varepsilon_0 r^2}$:

$$KE = \frac{1}{2}mv^2 = \frac{e^2}{8\pi\varepsilon_0 r}$$

Potential (Coulomb) energy is $PE = -\frac{e^2}{4\pi\varepsilon_0 r}$

PE is negative because electron and proton charges are of opposite sign.

$$E = PE + KE$$

Total energy $E = -\frac{e^2}{4\pi\varepsilon_0 r} + \frac{e^2}{8\pi\varepsilon_0 r} = -\frac{e^2}{8\pi\varepsilon_0 r}$

- The minus sign means energy is required to remove the electron

- This energy is called the binding energy.

Substituting for r from Equation $r = n^2 \left(\dfrac{\varepsilon_0 h^2}{\pi m e^2} \right) = n^2 a_0$ gives the energies of other allowed states:

$$E_n = -\frac{1}{n^2} \left(\frac{me^4}{8\varepsilon_0^2 h^2} \right) = -\frac{E_1}{n^2}$$

Lowest (ground) state of hydrogen: $E_1 = -13.6$ eV

Quantum Numbers and Energy Levels

Only certain orbit radii and total energies are allowed i.e. both electron radius and energy are quantized. n is called a quantum number

n	r	E
1	a_0	E_1
2	$4\,a_0$	$E_1/4$
3	$9\,a_0$	$E_1/9$
4	$16\,a_0$	$E_1/16$
--	--	--

Generalization to other single-electron atoms:

Replace `$e^2$` by `$Ze^2$`- e.g. singly ionized helium,

He^+ (Z = 2) or doubly ionized lithium, Li^{++} (Z = 3), etc.

$$r = n^2 \left(\frac{\varepsilon_0 h^2}{\pi m Z e^2} \right) = n^2 \frac{a_0}{Z}$$

and the energies $E_n = -\dfrac{Z^2}{n^2} E_1 = -13.6 \dfrac{Z^2}{n^2}$

These allowed energies are called energy levels.

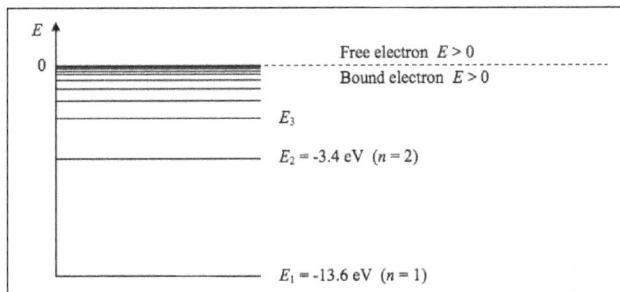

Allowed energy levels for the hydrogen atom.

- Normally, the electron is in the lowest state (n = 1), the ground state.

- It can gain energy from electromagnetic radiation, collisions with other atoms, etc. and be promoted into one of the higher levels, i.e. to an excited state of the atom.

- Note: the energy gained must equal the energy required.

- If the electron acquires energy > 13.6 eV, it is liberated from the atom altogether.

- The atom is ionized.

Energy Conservation and Spectral Lines

An electron in an excited state normally returns very quickly to its ground state, either directly or via an intermediate state.

- When the electron moves from its initial E_i to its final E_f state, a photon is emitted (or absorbed, if $E_i < E_f$).

- The energy of the photon emitted (or absorbed) is given by energy conservation:

$$E_{ph} = h\nu = E_i - E_f.$$

- For hydrogen: $E_{ph} = 13.6\left(\dfrac{1}{n_f^2} = \dfrac{1}{n_i^2}\right)\text{eV}$

where n_i and n_f (with ni > n_f) are the quantum numbers of the initial and final states.

Possible decays from an n = 3 excited atomic state:

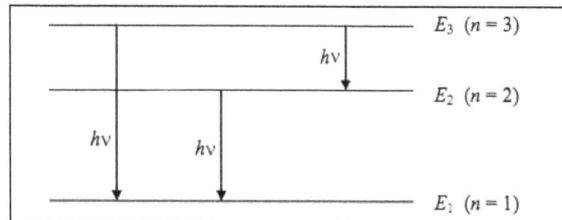

Possible decays to the ground state of a hydrogen atom: All these transitions correspond to discrete photon energies. A series of sharp spectral lines are produced. All these lines are in the ultraviolet region of the em spectrum. They are called the Lyman Series.

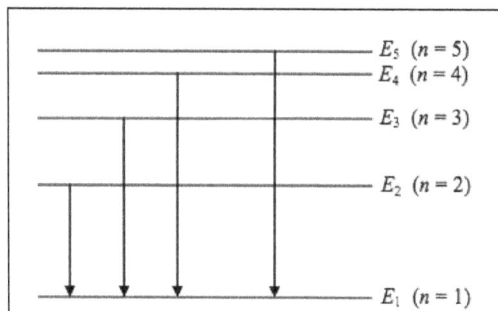

The Lyman Series

Initial quantum number	Photon energy (eV)	Wavelength $\lambda = hc/E$
2	$13.6 \times (1 - 1/4) = 10.2$	121.6 nm
3	$13.6 \times (1 - 1/9) = 12.1$	102.6 nm
4	$13.6 \times (1 - 1/16) = 12.75$	97.3 nm
5	$13.6 \times (1 - 1/25) = 13.1$	95.0 nm
---	---	---
∞	$13.6 \times (1 - 1/\infty) = 13.6$	91.2 nm

Electrons from states with $n_i > 2$ can return, initially, to the first-excited state ($n_f = 2$), emitting one photon and then to the ground state emitting a second photon, with $E_{ph} = 10.2$ eV, which is part of the Lyman series.

Photons from transitions to the first-excited, (n = 2) state of the hydrogen atom form another series of spectral lines. This series is in the visible part of the spectrum from yellow, for the lowest energy, to violet for the highest energies. It is called the Balmer Series.

Initial quantum number	Photon energy (eV)	Wavelength $\lambda = hc/E$
3	$13.6 \times (1/4 - 1/9) = 1.89$	657 nm
4	$13.6 \times (1/4 - 1/16) = 2.55$	487 nm
5	$13.6 \times (1/4 - 1/25) = 2.86$	434 nm
6	$13.6 \times (1/4 - 1/36) = 3.02$	411 nm
---	---	---
∞	$13.6 \times (1/4 - 1/\infty) = 3.4$	365 nm

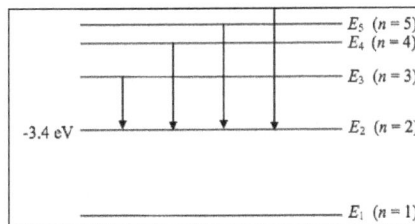

Other series of spectral lines correspond to transitions for which

$E_f = E_3$ (the Paschen series)

$E_f = E_4$ (the Brackett series) and so on.

These are all in the infra-red part of the spectrum.

Quantum Mechanical Model of Atom

Atomic model which is based on the particle and wave nature of the electron is known as wave or quantum mechanical model of the atom. This was developed by Erwin Schrodinger in 1926. This model describes the electron as a three dimensional wave in the electronic field of positively charged nucleus. Schrodinger derived an equation which describes wave motion of an electron. The differential equation is

$$\partial^2 \psi / \partial x^2 + \partial^2 \psi / \partial y^2 + \partial^2 \psi / \partial z^2 + 8\pi^2 m / h^2 \left(E - v \right) \psi = 0$$

where x, y, z are certain coordinates of the electron, m = mass of the electron E = total energy of the electron. V = potential energy of the electron; h = Planck's constant and ψ (psi) = wave function of the electron.

Significance of ψ: The wave function may be regarded as the amplitude function expressed in terms of coordinates x, y and z. The wave function may have positive or negative values depending upon the value of coordinates. The main aim of Schrodinger equation is to give solution for probability approach. When the equation is solved, it is observed that for some regions of space the value of ψ is negative. But the probability must be always positive and cannot be negative, it is thus, proper to use ψ^2 in favour of ψ.

Significance of ψ^2: ψ^2 is a probability factor. It describes the probability of finding an electron within a small space. The space in which there is maximum probability of finding an electron is termed as orbital. The important point of the solution of the wave equation is that it provides a set of numbers called quantum numbers which describe energies of the electron in atoms, information about the shapes and orientations of the most probable distribution of electrons around nucleus.

Nodal Points and Planes

The point where there is zero probability of finding the electron is called nodal point. There are two types of nodes: Radial nodes and angular nodes. The former is concerned with distance from the nucleus while latter is concerned with direction.

No. of radial nodes = n – l – 1

No. of angular nodes = l

Total number of nodes = n – 1

Nodal planes are the planes of zero probability of finding the electron. The number of such planes is also equal to l.

Quantum Numbers

An atom contains large number of shells and subshells (orbitals). These are distinguished from one another on the basis of their size, shape and orientation (direction) in space. The parameters are expressed in terms of different numbers called quantum numbers.

Quantum numbers may be defined as a set of four numbers with the help of which we can get complete information about all the electrons in an atom. It tells us the address of the electron i.e., location, energy, the type of orbital occupied and orientation of that orbital.

Principal quantum number (n): It tells the main shell in which the electron resides, the approximate distance of the electron from the nucleus and energy of that particular electron. It also tells the maximum number of electrons that a shell can accommodate is $2n^2$, where n is the principal quantum number.

Shell	K	L	M	N
Principal quantum number (n)	1	2	3	4
Maximum number of electrons	2	8	18	32

Azimuthal or angular momentum quantum number (l): This represents the number of subshells present in the main shell. These subsidiary orbits within a shell will be denoted as 0, 1, 2, 3, 4,... or s, p, d, f... This tells the shape of the subshells. The orbital angular momentum of the electron is given as $\sqrt{l(l+1)} h/2\pi$ or $\sqrt{l(l+1)} \hbar$ for a particular value of 'n' (where $\hbar = h/2\pi$). For a given value of n values of possible l vary from 0 to n − 1.

The magnetic quantum number (m): An electron due to its angular motion around the nucleus generates an electric field. This electric field is expected to produce a magnetic field. Under the influence of external magnetic field, the electrons of a subshell can orient themselves in certain preferred regions of space around the nucleus called orbitals. The magnetic quantum number determines the number of preferred orientations of the electron present in a subshell. The values allowed depends on the value of l, the angular momentum quantum number, m can assume all integral values between −l to +l including zero. Thus m can be −1, 0, +1 for l = 1.Total values of m associated with a particular value of l is given by 2l+ 1.

The spin quantum number (s): Just like earth which not only revolves around the sun but also spins about its own axis, an electron in an atom not only revolves around the nucleus but also spins about its own axis. Since an electron can spin either in clockwise direction or in anticlockwise direction, therefore, for any particular value of magnetic quantum number, spin quantum number can have two values, i.e., +1/2 and −1/2 or these are represented by two arrows pointing in the opposite directions, i.e., ↑ and ↓. When an electron goes to a vacant orbital, it can have a clockwise or anti clockwise spin i.e., +1/2 or −1/2. This quantum number helps to explain the magnetic properties of the substances.

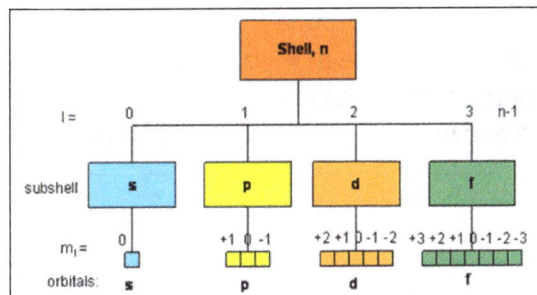

Quantum Number	Symbol	Values
Principal	n	1,2,...
Angular Momentum	l	0,1,2,...,n−1
Magnetic	m	−l to +l
Spin Magnetic	s	+1/2,−1/2

Shapes and Size of Orbitals

An orbital is the region of space around the nucleus within which the probability of finding an electron of given energy is maximum (90–95%). The shape of this region (electron cloud) gives the shape of the orbital. It is basically determined by the azimuthal quantum number l , while the orientation of orbital depends on the magnetic quantum number (m).

s–orbital:These orbitals are spherical and symmetrical about the nucleus. The probability of finding the electron is maximum near the nucleus and keeps on decreasing as the distance from the nucleus increases. There is vacant space between two successive s–orbitals known as radial node. But there is no radial node for 1s orbital since it is starting from the nucleus.

p–orbital (l=1):The size of the orbital depends upon the value of principal quantum number (n). Greater the value of n, larger is the size of the orbital. Therefore, 2s–orbital is larger than 1s orbital but both of them are non-directional and spherically symmetrical in shape.

The probability of finding the p–electron is maximum in two lobes on the opposite sides of the nucleus. This gives rise to a dumb–bell shape for the p–orbital. For p–orbital l = 1. Hence, m = −1, 0, +1. Thus, p–orbital have three different orientations. These are designated as px, py & pz depending upon whether the density of electron is maximum along the x y and z axis respectively. As they are not spherically symmetrical, they have directional character. The two lobes of p–orbitals are separated by a nodal plane, where the probability of finding electron is zero.

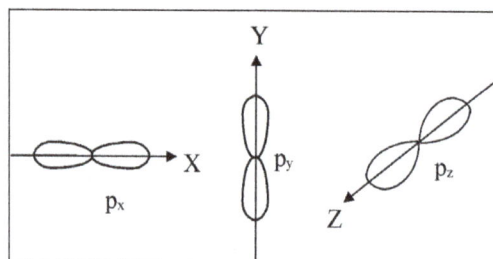

The three p-orbitals belonging to a particular energy shell have equal energies and are called degenerate orbitals.

d–orbital (= 2): For d–orbitals, l = 2. Hence m = −2,−1, 0, +1, +2. Thus there are 5d orbitals. They have relatively complex geometry. Out of the five orbitals, the three (dxy, dyz, dzx) project in between the axis and the other two dz2 and $d_{z^2-y^2}$ lie along the axis.

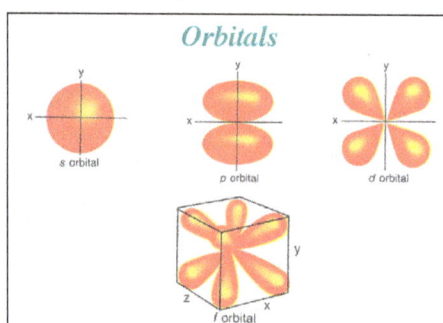

Orbitals

Aufbau Principle

The Aufbau principle dictates the manner in which electrons are filled in the atomic orbitals of an atom in its ground state. It states that electrons are filled into atomic orbitals in the increasing order of orbital energy level. According to the Aufbau principle, the available atomic orbitals with the lowest energy levels are occupied before those with higher energy levels.

The word 'Aufbau' has German roots and can be roughly translated as 'construct' or 'build up'. A diagram illustrating the order in which atomic orbitals are filled is provided below. Here, 'n' refers to the principal quantum number and 'l' is the azimuthal quantum number.

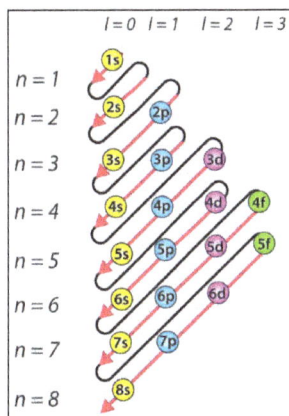

The Aufbau principle can be used to understand the location of electrons in an atom and their corresponding energy levels. For example, carbon has 6 electrons and its electronic configuration is $1s^2 2s^2 2p^2$.

It is important to note that each orbital can hold a maximum of two electrons (as per the Pauli exclusion principle). Also, the manner in which electrons are filled into orbitals in a single subshell must follow Hund's rule, i.e. every orbital in a given subshell must be singly occupied by electrons before any two electrons pair up in an orbital.

Salient Features of the Aufbau Principle

- According to the Aufbau principle, electrons first occupy those orbitals whose energy is the lowest. This implies that the electrons enter the orbitals having higher energies only when orbitals with lower energies have been completely filled.

- The order in which the energy of orbitals increases can be determined with the help of the (n+l) rule, where the sum of the principal and azimuthal quantum numbers determines the energy level of the orbital.

- Lower (n+l) values correspond to lower orbital energies. If two orbitals share equal (n+l) values, the orbital with the lower n value is said to have lower energy associated with it.

- The order in which the orbitals are filled with electrons is: 1s, 2s, 2p, 3s, 3p, 4s, 3d, 4p, 5s, 4d, 5p, 6s, 4f, 5d, 6p, 7s, 5f, 6d, 7p, and so on.

Exceptions

The electron configuration of chromium is $[Ar]3d^54s^1$ and not $[Ar]3d^44s^2$ (as suggested by the Aufbau principle). This exception is attributed to several factors such as the increased stability provided by half-filled subshells and the relatively low energy gap between the 3d and the 4s subshells.

The energy gap between the different subshells is illustrated below.

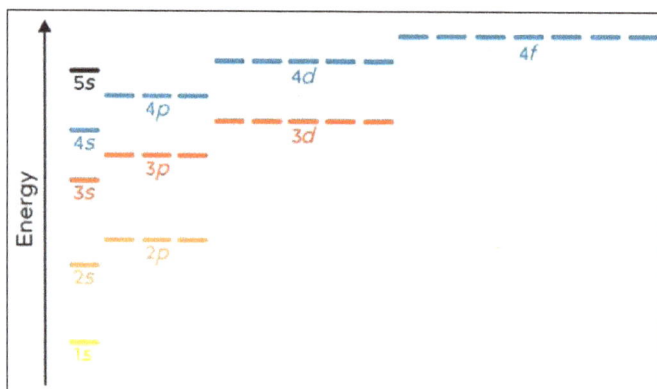

Half filled subshells feature lower electron-electron repulsions in the orbitals, thereby increasing the stability. Similarly, completely filled subshells also increase the stability of the atom. Therefore, the electron configurations of some atoms disobey the Aufbau principle (depending on the energy gap between the orbitals).

For example, copper is another exception to this principle with an electronic configuration corresponding to $[Ar]3d^{10}4s^1$. This can be explained by the stability provided by a completely filled 3d subshell.

Electronic Configuration using the Aufbau Principle

Writing the Electron Configuration of Sulfur

- The atomic number of sulfur is 16, implying that it holds a total of 16 electrons.

- As per the Aufbau principle, two of these electrons are present in the 1s subshell, eight of them are present in the 2s and 2p subshell, and the remaining are distributed into the 3s and 3p subshells.

- Therefore, the electron configuration of sulfur can be written as $1s^22s^22p^63s^23p^6$.

Writing the Electron Configuration of Nitrogen

- The element nitrogen has 7 electrons (since its atomic number is 7).

- The electrons are filled into the 1s, 2s, and 2p orbitals.

- The electron configuration of nitrogen can be written as $1s^22s^22p^3$.

Hund's Rules

Aufbau principle tells us that the lowest energy orbitals get filled by electrons first. After the lower energy orbitals are filled, the electrons move on to higher energy orbitals. The problem with this rule is that it does not tell about the three 2p orbitals and the order that they will be filled in. According to Hund's rule:

- Before the double occupation of any orbital, every orbital in the sub level is singly occupied.

- For the maximization of total spin, all electrons in a single occupancy orbital have the same spin.

An electron will not pair with another electron in a half-filled orbital as it has the ability to fill all its orbitals with similar energy. A large number of unpaired electrons are present in atoms which are at the ground state. If two electrons come in contact they would show the same behaviour as two magnets do. The electrons first try to get as far away from each other as possible before they have to pair up.

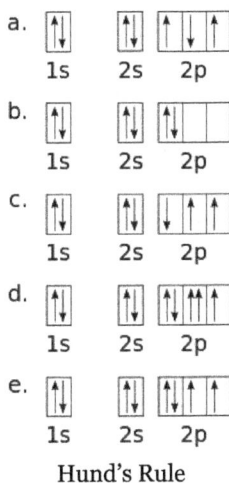

Hund's Rule

It states that:

1. In a sublevel, each orbital is singly occupied before it is doubly occupied.

2. The electrons present in singly occupied orbitals possess identical spin.

Explanation of Hund's Rule

The electrons enter an empty orbital before pairing up. The electrons repel each other as they are negatively charged. The electrons do not share orbitals to reduce repulsion.

When we consider the second rule, the spins of unpaired electrons in singly occupied orbitals is the same. The initial electrons spin in the sub level decides what the spin of the other electrons would be. For instance, a carbon atom's electron configuration would be $1s^22s^22p^2$. The same orbital will be occupied by the two 2s electrons although different orbitals will be occupied by the two 2p electrons in reference to Hund's rule.

Electron Configuration and its Purpose

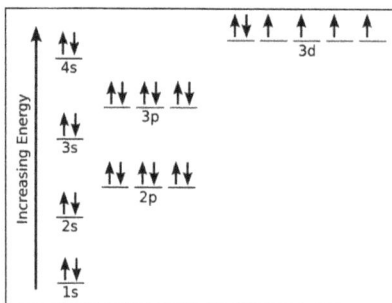

Electron Configuration

The above image helps in understanding the electronic configuration and its purpose. The valence shells of two atoms that come in contact with each other will interact first. When valence shells are not full then the atom is least stable. The chemical characteristics of an element are largely dependent on the valence electrons. Similar chemical characteristics can be seen in elements that have similar valence numbers.

The stability can also be predicted by the electron configuration. When all the orbitals of an atom are full it is most stable. The orbitals that have full energy level are the most stable, for example, noble gases. These type of elements do not react with other elements.

Hund's Rule of Maximum Multiplicity

The rule states that, for a stated electron configuration, the greatest value of spin multiplicity has the lowest energy term. It says if two or more than two orbitals having the same amount of energy are unoccupied then the electrons will start occupying them individually before they fill them in pairs. It is a rule which depends on the observation of atomic spectra, that is helpful in predicting the ground state of a molecule or an atom with one or more than one open electronic shells. This rule was discovered in the year 1925 by Friedrich Hund.

Uses of hund's rule: It has wide applications, for example, it is majorly used in atomic chemistry, quantum chemistry, and spectroscopy, etc.

Pauli Exclusion Principle

The Pauli Exclusion Principle states that, in an atom or molecule, no two electrons can have the same four electronic quantum numbers. As an orbital can contain a maximum of only two electrons, the two electrons must have opposing spins. This means if one is assigned an up-spin (+1/2), the other must be down-spin (-1/2).

Electrons in the same orbital have the same first three quantum numbers, e.g., $n=1$, $l=0$, $m_i =0$ for the 1s subshell. Only two electrons can have these numbers, so that their spin moments must be either $m_s = -1/2 \, or \, m_s = +1/2$. If the 1s orbital contains only one electron, we have one ms value and the electron configuration is written as $1s^1$ (corresponding to hydrogen). If it is fully occupied, we have two m_s values, and the electron configuration is $1s^2$ (corresponding to helium). Visually these two cases can be represented as

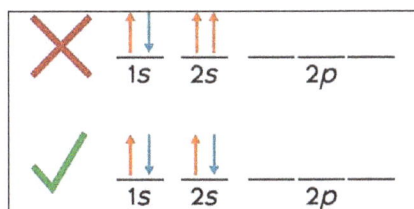

As you can see, the 1s and 2s subshells for beryllium atoms can hold only two electrons and when filled, the electrons must have opposite spins. Otherwise they will have the same four quantum numbers, in violation of the Pauli Exclusion Principle.

Pauli Exclusion Principle

Consider argon's electron configuration:

$$1s^2 \ 2s^2 \ 2p^6 \ 3s^2 \ 3p^6$$

The exclusion principle asserts that every electron in an argon atom is in a unique state.

The 1s level can accommodate two electrons with identical n, l, and ml quantum numbers. Argon's pair of electrons in the 1s orbital satisfy the exclusion principle because they have opposite spins, meaning they have different spin quantum numbers, m_s. One spin is $+\frac{1}{2}$, the other is $-\frac{1}{2}$. (Instead of saying $+\frac{1}{2}$ or $-\frac{1}{2}$ often the electrons are said to be spin-up up arrow or spin-down down arrow.)

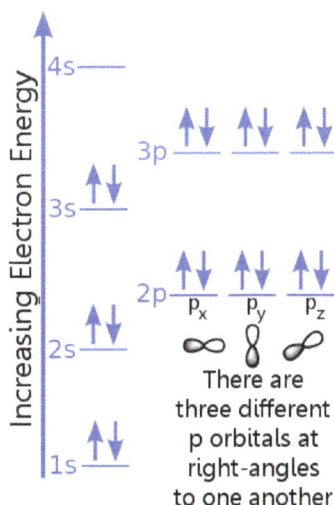

The 2s level electrons have a different principal quantum number to those in the 1s orbital. The pair of 2s electrons differ from each other because they have opposite spins.

The 2p level electrons have a different orbital angular momentum number from those in the s orbitals, hence the letter p rather than s. There are three p orbitals of equal energy, the px, py and pz. These orbitals are different from one another because they have different orientations in space. Each of the px, py and pz orbitals can accommodate a pair of electrons with opposite spins.

The 3s level rises to a higher principal quantum number; this orbital accommodates an electron pair with opposite spins.

The 3p level's description is similar to that for 2p, but the principal quantum number is higher: 3p lies at a higher energy than 2p.

Reference

- Atomic-physics, science: britannica.com, Retrieved 10 February, 2019

- Atom-definition: livescience.com, Retrieved 15 May, 2019

- Atomic-models, structure-of-atom, chemistry: toppr.com, Retrieved 13 April, 2019

- Bohr-model-of-the-atom: thoughtco.com, Retrieved 17 July, 2019

- Quantum-mechanical-model-of-atom, atomic-structure: askiitians.com, Retrieved 19 January, 2019

- Aufbau-principle, chemistry: byjus.com, Retrieved 22 May, 2019

- Hunds-rule, chemistry: byjus.com, Retrieved 18 June, 2019

- Pauli-exclusion-principle: chemicool.com, Retrieved 5 March, 2019

10

Nuclear Physics

Nuclear physics is a domain of physics that is concerned with the study of atomic nuclei, their constituents and interactions. It is applied in various fields such as nuclear power, nuclear medicine, magnetic resonance imaging, nuclear weapons, industrial and agricultural isotopes, etc. The aim of this chapter is to provide an easy understanding of the varied facets of nuclear physics.

Nuclear physics is the field of physics that studies atomic nuclei. In other words, nuclear physics deals with the components and structure of the nucleus. Nuclear reaction comprises of the merging of nuclei, radioactive decay, fusion, fission and break-up of a nucleus.

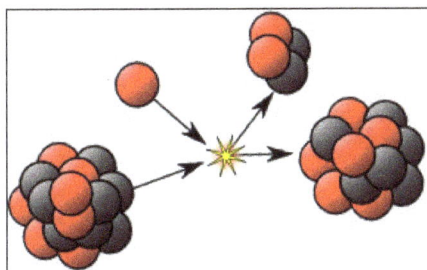

Nuclear Physics and Atomic Physics

- The simple difference between nuclear physics and atomic physics is that nuclear physics deals with the nucleus while atomic physics deals with an entire atom. But isn't nucleus a part of the atom? Then why do we have two separate branches?

- Atomic physics deals with the properties of an atom as a whole, mainly due to its electronic configuration. Of course, nucleus is a part of this but only in terms of its overall contribution.

- Nuclear physics, on the other hand, deals exclusively with nuclei, their structures, properties, reactions and interactions. Nuclear Physics Atoms make up all the matter in the universe.

- Nucleus is densely concentrated at the core of an atom. The study and research on nuclear physics concerns itself with the forces that are able to hold this dense matter at the core among other things like understanding quarks and gluons.

- The application of nuclear physics is largely in the field of power generation using nuclear energy. Once the force holding the nucleus was understood, we started splitting and fusing neutrons.

- The energy evolved in this process can be used in the splitting of the nucleus to generate energy in Nuclear Fission and fusing two neutrons to generate energy is Nuclear Fusion.

Radius of Nucleus

'R' represents the radius of nucleus.

$$R = R_o \, A^{\frac{1}{3}}$$

Here,

Ro = Proportionality Constant

A = mass number of the element.

Total Number of Protons and Neutrons in a Nucleus

The Mass number (A), also known as nucleon number, is the total number of neutrons and protons in a nucleus.

$$A = Z + N$$

Where,

N = Neutron number

A = Mass number

Z = Proton number

Mass Defect

When nuclei are formed, some of the mass gets lost during the process and this lost mass is known as a mass defect.

$$\Delta m = Z m_p + (A - Z) m_n - M$$

Here,

M = Mass of the nucleus

Δm = Mass of the nucleons – Mass of the nucleus

m_p = Mass of the Proton

m_n = Mass of the Neutron

Packing Fraction:

Packing fraction is defined as Mass defect per nucleon.

Packing fraction (f) = Mass defect per nucleons

$$\text{Packing fraction}(f) = \frac{[Z m_p + (A - Z) m_n - A]}{A}$$

Radioactivity

Radioactive decay is the process by which an excited, unstable atomic nucleus loses energy by emitting radiation in the form of particles or electromagnetic waves, thereby transitioning toward a more stable state.

The atomic nucleus comprises certain combinations of protons and neutrons held in a stable configuration through a precise balance of powerful forces: The strong force holding the protons and neutrons together is powerful but very short range; the electrostatic repulsion of the positively charged protons is less powerful but long range; the weak force makes the neutron inherently unstable and will turn it into a proton if given the chance. This balance is very delicate: a uranium-238 nucleus has a half-life of 4.5 billion years while uranium-237 with just one less neutron has a half-life of 1.3 minutes.

If there is an imbalance in these forces, the system will eventually shed the excess by ejecting radiation in some combination of particles and wave energy. The most common radioactive decays occur in response to one of three possible types of imbalance. If the nucleus has too many neutrons, one of its neutrons decays (through beta decay) into one proton plus two fragments ejected from the nucleus, a neutrino and an electron (called a a beta particle). If the nucleus has too many protons, it undergoes alpha decay by ejecting two protons and two neutrons as an alpha particle. If the nucleus is excited (has too much energy) it ejects a gamma ray.

Materials exhibiting radioactive decay have yielded widespread application to enhance human welfare. The various applications take advantage of the different decay properties, different decay products, and different chemical properties of the many elements having some isotopes that are radioactive. Major types of applications use the radiation either for diagnosing a problem or for treating a problem by killing specific harmful cells. Areas of application include human and veterinary medicine, nutrition research, basic research in genetics and metabolism, household smoke detectors, industrial and mining inspection of welds, security inspection of cargo, tracing and analyzing pollutants in studies of runoff, and dating materials in geology, paleontology, and archaeology.

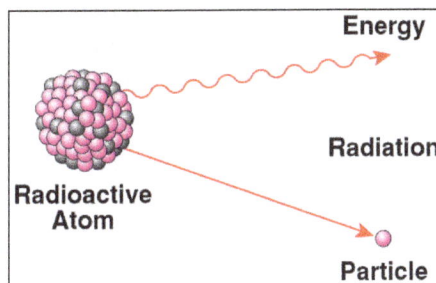

Radioactive decay modes of an atomic nucleus.

Nuclides

Radioactive decay results in an atom of one type, called the *parent nuclide,* being transformed to an atom of a different type, called the *daughter nuclide.* For example, a carbon-14 atom (the "parent") emits radiation and transforms to a nitrogen-14 atom (the "daughter"). This transformation involves quantum probability, so it is impossible to predict when a particular atom will decay. Given a large number of atoms, however, the decay rate is predictable and measured by the "half-life"—the time it takes for 50 percent of the atoms to undergo the change. The half-life of radioactive atoms varies enormously; from fractions of a millisecond to billions of years.

The trefoil symbol is used to indicate radioactive material.

The SI unit of radioactive decay (the phenomenon of natural and artificial radioactivity) is the becquerel (Bq). One Bq is defined as one transformation (or decay) per second. Since any reasonably-sized sample of radioactive material contains many atoms, a Bq is a tiny measure of activity; amounts on the order of TBq (terabecquerel) or GBq (gigabecquerel) are commonly used. Another unit of (radio)activity is the curie, Ci, which was originally defined as the activity of one gram of pure radium, isotope Ra-226. At present, it is equal (by definition) to the activity of any radionuclide decaying with a disintegration rate of 3.7×10^{10} Bq. The use of Ci is presently discouraged by SI.

The neutrons and protons that constitute nuclei, as well as other particles that may approach them, are governed by several interactions. The strong nuclear force, not observed at the familiar macroscopic scale, is the most powerful force over subatomic distances. The electrostatic force is also significant, while the weak nuclear force is responsible for Beta decay.

The interplay of these forces is simple. Some configurations of the particles in a nucleus have the property that, should they shift ever so slightly, the particles could fall into a lower-energy arrangement (with the extra energy moving elsewhere). One might draw an analogy with a snowfield on a mountain: While friction between the snow crystals can support the snow's weight, the system is inherently unstable with regards to a lower-potential-energy state, and a disturbance may facilitate the path to a greater entropy state (that is, towards the ground state where heat will be produced, and thus total energy is distributed over a larger number of quantum states). Thus, an avalanche results. The total energy does not change in this process, but because of entropy effects, avalanches only happen in one direction, and the end of this direction, which is dictated by the largest number of chance-mediated ways to distribute available energy, is what we commonly refer to as the "ground state."

Such a collapse (a decay event) requires a specific activation energy. In the case of a snow avalanche, this energy classically comes as a disturbance from outside the system, although such disturbances can be arbitrarily small. In the case of an excited atomic nucleus, the arbitrarily small disturbance comes from quantum vacuum fluctuations. A nucleus (or any excited system in quantum mechanics) is unstable, and can thus spontaneously stabilize to a less-excited system. This process is driven by entropy considerations: The energy does not change, but at the end of the process, the total energy is more diffused in spacial volume. The resulting transformation alters the structure of the nucleus. Such a reaction is thus a nuclear reaction, in contrast to chemical reactions, which also are driven by entropy, but which involve changes in the arrangement of the outer electrons of atoms, rather than their nuclei.

Some nuclear reactions do involve external sources of energy, in the form of collisions with outside particles. However, these are not considered decay. Rather, they are examples of induced nuclear reactions. Nuclear fission and fusion are common types of induced nuclear reactions.

Discovery

Radioactivity was first discovered in 1896, by the French scientist Henri Becquerel while working on phosphorescent materials. These materials glow in the dark after exposure to light, and he thought that the glow produced in cathode ray tubes by X-rays might somehow be connected with phosphorescence. So, he tried wrapping a photographic plate in black paper and placing various phosphorescent minerals on it. All results were negative until he tried using uranium salts. The result with these compounds was a deep blackening of the plate.

However, it soon became clear that the blackening of the plate had nothing to do with phosphorescence because the plate blackened when the mineral was kept in the dark. Also, non-phosphorescent salts of uranium and even metallic uranium blackened the plate. Clearly there was some new form of radiation that could pass through paper that was causing the plate to blacken.

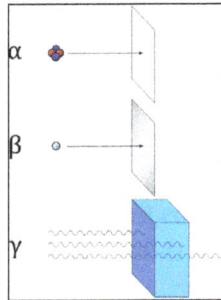

Alpha particles may be completely stopped by a sheet of paper, beta particles by aluminum shielding. Gamma rays, however, can only be reduced by much more substantial obstacles, such as a very thick piece of lead.

At first, it seemed that the new radiation was similar to the then recently discovered X-rays. However, further research by Becquerel, Marie Curie, Pierre Curie, Ernest Rutherford, and others discovered that radioactivity was significantly more complicated. Different types of decay can occur, but Rutherford was the first to realize that they all occur with the same mathematical, approximately exponential, formula.

As for types of radioactive radiation, it was found that an electric or magnetic field could split such emissions into three types of beams. For lack of better terms, the rays were given the alphabetic names alpha, beta, and gamma; names they still hold today. It was immediately obvious from the direction of electromagnetic forces that alpha rays carried a positive charge, beta rays carried a negative charge, and gamma rays were neutral. From the magnitude of deflection, it was also clear that alpha particles were much more massive than beta particles. Passing alpha rays through a thin glass membrane and trapping them in a discharge tube allowed researchers to study the emission spectrum of the resulting gas, and ultimately prove that alpha particles are in fact helium nuclei. Other experiments showed the similarity between beta radiation and cathode rays; they are both streams of electrons, and between gamma radiation and X-rays, which are both high energy electromagnetic radiation.

Although alpha, beta, and gamma are most common, other types of decay were eventually discovered. Shortly after discovery of the neutron in 1932, it was discovered by Enrico Fermi that certain rare decay reactions give rise to neutrons as a decay particle. Isolated proton emission was also eventually observed in some elements. Shortly after the discovery of the positron in cosmic ray products, it was realized that the same process that operates in classical beta decay can also produce positrons (positron emission), analogously to negative electrons. Each of the two types of beta decay acts to move a nucleus toward a ratio of neutrons and protons which has the least energy for the combination. Finally, in a phenomenon called cluster decay, specific combinations of neutrons and protons other than alpha particles were found to occasionally spontaneously be emitted from atoms.

Still other types of radioactive decay were found which emit previously seen particles, but by different mechanisms. An example is internal conversion, which results in electron and sometimes high energy photon emission, even though it involves neither beta nor gamma decay.

The early researchers also discovered that many other chemical elements besides uranium have radioactive isotopes. A systematic search for the total radioactivity in uranium ores also guided Marie Curie to isolate a new element, polonium, and to separate a new element, radium, from barium; the two elements' chemical similarity would otherwise have made them difficult to distinguish.

The dangers of radioactivity and of radiation were not immediately recognized. Acute effects of radiation were first observed in the use of X-rays when the Serbo-Croatian-American electric engineer, Nikola Tesla, intentionally subjected his fingers to X-rays in 1896. He published his observations concerning the burns that developed, though he attributed them to ozone rather than to the X-rays. Fortunately, his injuries healed later.

The genetic effects of radiation, including the effects on cancer risk, were recognized much later. It was only in 1927 that Hermann Joseph Muller published his research that showed the genetic effects. In 1946, he was awarded the Nobel prize for his findings.

Before the biological effects of radiation were known, many physicians and corporations had begun marketing radioactive substances as patent medicine, much of which was harmful to health and gave rise to the term radioactive quackery; particularly alarming examples were radium enema treatments, and radium-containing waters to be drunk as tonics. Marie Curie spoke out against this sort of treatment, warning that the effects of radiation on the human body were not well understood (Curie later died from aplastic anemia, assumed due to her own work with radium, but later examination of her bones showed that she had been a careful laboratory worker and had a low burden of radium; a better candidate for her disease was her long exposure to unshielded X-ray tubes while a volunteer medical worker in World War I). By the 1930s, after a number of cases of bone-necrosis and death in enthusiasts, radium-containing medical products had nearly vanished from the market.

Modes of Decay

Radionuclides can undergo a number of different reactions. These are summarized in the following table. A nucleus with atomic weight A and a positive charge Z (called atomic number) is represented as (A, Z).

Mode of decay	Participating particles	Daughter nucleus
Decays with emission of nucleons:		
Alpha decay	An alpha particle (A=4, Z=2) emitted from nucleus	(A-4, Z-2)
Proton emission	A proton ejected from nucleus	(A-1, Z-1)
Neutron emission	A neutron ejected from nucleus	(A-1, Z)
Double proton emission	Two protons ejected from nucleus simultaneously	(A-2, Z-2)
Spontaneous fission	Nucleus disintegrates into two or more smaller nuclei and other particles	-
Cluster decay	Nucleus emits a specific type of smaller nucleus (A_1, Z_1) larger than an alpha particle	(A-A_1, Z-Z_1) + (A_1,Z_1)
Different modes of beta decay:		
Beta-Negative decay	A nucleus emits an electron and an antineutrino	(A, Z+1)
Positron emission, also Beta-Positive decay	A nucleus emits a positron and a neutrino	(A, Z-1)
Electron capture	A nucleus captures an orbiting electron and emits a neutrino - The daughter nucleus is left in an excited and unstable state	(A, Z-1)
Double beta decay	A nucleus emits two electrons and two antineutrinos	(A, Z+2)
Double electron capture	A nucleus absorbs two orbital electrons and emits two neutrinos - The daughter nucleus is left in an excited and unstable state	(A, Z-2)
Electron capture with positron emission	A nucleus absorbs one orbital electron, emits one positron and two neutrinos	(A, Z-2)
Double positron emission	A nucleus emits two positrons and two neutrinos	(A, Z-2)
Transitions between states of the same nucleus:		
Gamma decay	Excited nucleus releases a high-energy photon (gamma ray)	(A, Z)
Internal conversion	Excited nucleus transfers energy to an orbital electron and it is ejected from the atom	(A, Z)

Radioactive decay results in a reduction of summed rest mass, which is converted to energy (the *disintegration energy*) according to the formula $E = mc^2$. This energy is released as kinetic energy of the emitted particles. The energy remains associated with a measure of mass of the decay system invariant mass, inasmuch the kinetic energy of emitted particles contributes also to the total invariant mass of systems. Thus, the sum of rest masses of particles is not conserved in decay, but the *system* mass or system invariant mass (as also system total energy) is conserved.

Radioactive Series

In a simple, one-step radioactive decay, the new nucleus that emerges is stable. C-14 undergoing beta decay to N-14 and K-40 undergoing electron capture to Ar-40 are examples.

On the other hand, the daughter nuclide of a decay event can be unstable, sometimes even more unstable than the parent. If this is the case, it will proceed to decay again. A sequence of several decay events, producing in the end a stable nuclide, is a *decay chain*. Ultrapure uranium, for instance, is hardly radioactive at all. After a few weeks, however, the unstable daughter nuclides accumulate—such as radium—and it is their radioactivity that becomes noticeable.

Of the commonly occurring forms of radioactive decay, the only one that changes the number of aggregate protons and neutrons *(nucleons)* contained in the nucleus is alpha emission, which reduces it by four. Thus, the number of nucleons modulo 4 is preserved across any decay chain. This leads to the four radioactive decay series with atomic weights 4n+0, 4n+1, 4n+2, and 4n+3.

In an alpha decay, the atomic weight decreases by 4 and the atomic number decreases by 2. In a beta decay, the atomic weight stays the same and the atomic number increases by 1. In a gamma decay, both atomic weight and number stay the same. A branching path occurs when there are alternate routes to the same stable destination. One branch is usually highly favored over the other.

These are the four radioactive decay series:

Uranium-235 Series (4n+3)

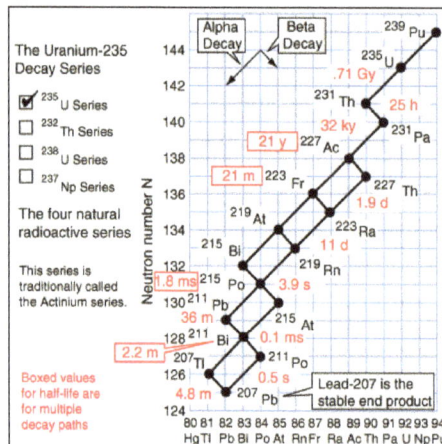

The Uranium-235 Decay Series. Alpha Decay / Beta Decay. ^{235}U Series, ^{232}Th Series, ^{238}U Series, ^{237}Np Series. The four natural radioactive series. This series is traditionally called the Actinium series. Boxed values for half-life are for multiple decay paths. Neutron number N. Nuclides shown include ^{239}Pu, ^{235}U (.71 Gy), ^{231}Th (25 h), ^{231}Pa, ^{227}Ac (32 ky, 21 y), ^{227}Th (1.9 d), ^{223}Fr (21 m), ^{223}Ra (11 d), ^{219}At, ^{219}Rn (3.9 s), ^{215}Bi, ^{215}Po (1.8 ms), ^{215}At (0.1 ms), ^{211}Pb (36 m), ^{211}Bi (2.2 m), ^{211}Po (0.5 s), ^{207}Tl (4.8 m), ^{207}Pb. Lead-207 is the stable end product.

Thorium-232 Series (4n+0)

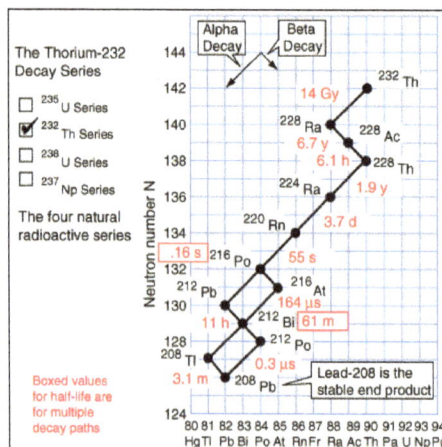

The Thorium-232 Decay Series. Alpha Decay / Beta Decay. ^{235}U Series, ^{232}Th Series, ^{238}U Series, ^{237}Np Series. The four natural radioactive series. Boxed values for half-life are for multiple decay paths. Neutron number N. Nuclides shown include ^{232}Th (14 Gy), ^{228}Ra (6.7 y), ^{228}Ac (6.1 h), ^{228}Th (1.9 y), ^{224}Ra (3.7 d), ^{220}Rn (55 s), ^{216}Po (.16 s), ^{216}At (164 μs), ^{212}Pb (11 h), ^{212}Bi (61 m), ^{212}Po (0.3 μs), ^{208}Tl (3.1 m), ^{208}Pb. Lead-208 is the stable end product.

Uranium-238 Series (4n+2)

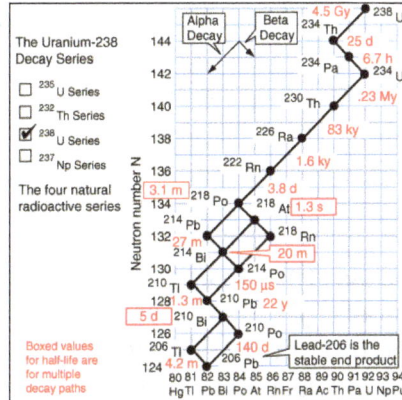

The Uranium-238 Decay Series

- ☐ ^{235}U Series
- ☐ ^{232}Th Series
- ☑ ^{238}U Series
- ☐ ^{237}Np Series

The four natural radioactive series

Boxed values for half-life are for multiple decay paths

Neptunium-237 Series (4n+1)

The members of this series are not presently found in nature because the half-life of the longest lived isotope in the series is short compared to the age of the earth.

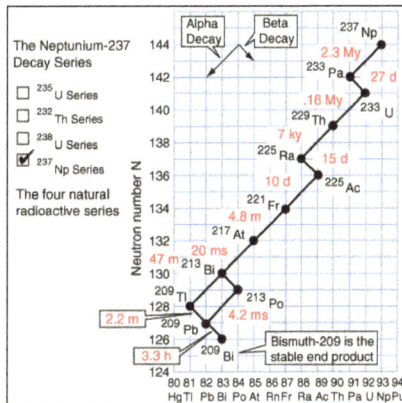

The Neptunium-237 Decay Series

- ☐ ^{235}U Series
- ☐ ^{232}Th Series
- ☐ ^{238}U Series
- ☑ ^{237}Np Series

The four natural radioactive series

Occurrence

According to the widely accepted Big Bang theory, the universe began as a mixture of hydrogen-1 (75 percent) and helium-4 (25 percent) with only traces of other light atoms. All the other elements, including the radioactive ones, were generated later during the thermonuclear burning of stars—the fusion of the lighter elements into the heavier ones. Stable isotopes of the lightest five elements (H, He, and traces of Li, Be, and B) were produced very shortly after the emergence of the universe, in a process called Big Bang nucleosynthesis. These lightest stable nuclides (including deuterium) survive to today, but any radioactive isotopes of the light elements produced in the Big Bang (such as tritium) have long since decayed. Isotopes of elements heavier than boron were not produced at all in the Big Bang, and these first five elements do not have any long-lived radioisotopes. Thus, all radioactive nuclei are, therefore, relatively young with respect to the birth of the universe, having formed later in various other types of nucleosynthesis in stars (in particular, supernovae), and also during ongoing interactions between stable isotopes and energetic particles. For example, carbon-14, a radioactive nuclide with a half-life of only 5,730 years, is constantly produced in Earth's upper atmosphere due to interactions between cosmic rays and nitrogen.

Applications

Radioactive materials and their decay products—alpha particles (2 protons plus 2 neutrons), beta particles (electrons or positrons), gamma radiation, and the daughter isotopes—have been put to the service of humanity in a great number of ways. At the same time, high doses of radiation from radioactive materials can be toxic unless they are applied with medical precision and control. Such exposures are unlikely except for the unlikely cases of a nuclear weapon detonation or an accident or attack on a nuclear facility.

In medicine, some radioactive isotopes, such as iron-59 and iodine-131, are usable directly in the body because the isotopes are chemically the same as stable iron and iodine respectively. Iron-59, steadily announcing its location by emitting beta-decay electrons, is readily incorporated into blood cells and thereby serves as an aid in studying iron deficiency, a nutritional deficiency affecting more than 2 billion people globally. Iron-59 is an important tool in the effort to understand the many factors affecting a person's ability to metabolize iron in the diet so that it becomes part of the the blood. Iodine-131 administered in the blood to people suffering from hyperthyroidism or thyroid cancer concentrates in the thyroid where gamma radiation emitted by the iodine-131 kills many of the thyroid cells. Hyperthyroidism in cats is treated effectively by one dose of iodine-131.

Radioactive isotopes whose chemical nature does not permit them to be readily incorporated into the body, are delivered to targeted areas by attaching them to a particular molecule that does tend to concentrate in a particular bodily location—just as iodine naturally concentrates in the thyroid gland. For studying activity in the brain, the radioactive isotope fluorine-18 is commonly attached to an analog of the sugar glucose which tends to concentrate in the active regions of the brain within a short time after the molecule is injected into the blood. Fluorine-18 decays by releasing a positron whose life is soon ended as it meets an electron and the two annihilate yielding gamma radiation that is readily detected by the Positron Emission Tomography (PET) technology. Similar techniques of radioisotopic labeling, have been used to track the passage of a variety of chemical substances through complex systems, especially living organisms.

Three gamma emitting radioisotopes are commonly used as a source of radiation. Technetium-99m, a metastable form with a half-life of 6 hours, emits a relatively low frequency gamma radiation that is readily detected. It has been widely used for imaging and functional studies of the brain, myocardium, thyroid, lungs, liver, gallbladder, kidneys, skeleton, blood, and tumors. Gamma radiation from cobalt-60 is used for sterilizing medical equipment, treating cancer, pasteurizing certain foods and spices, gauging the thickness of steel as it is being produced, and monitoring welds. Cesium-137 is used as a source of gamma radiation for treating cancer, measuring soil density at construction sites, monitoring the filling of packages of foods and pharmaceuticals, monitoring fluid flows in production plants, and studying rock layers in oil wells.

Americanium-241, which decays by emitting alpha particles and low energy gamma radiation, is commonly used in smoke detectors as the alpha particles ionize air in a chamber permitting a small current to flow. Smoke particles entering the chamber activate the detector by absorbing alpha particles without being ionized, thereby reducing the current.

On the premise that radioactive decay is truly random (rather than merely chaotic), it has been used in hardware random-number generators. Because the process is not thought to vary significantly in mechanism over time, it is also a valuable tool in estimating the absolute ages of certain materials.

For geological materials, the radioisotopes (parents) and certain of their decay products (daughters) become trapped when a rock solidifies, and can then later be used to estimate the date of the solidification (subject to such uncertainties as the possible number of daughter elements present at the time of solidification and the possible number of parent or daughter atoms added or removed over time).

For dating organic matter, radioactive carbon-14 is used because the atmosphere contains a small percentage of carbon-14 along with the predominance of stable carbons 12 and 13. Living plants incorporate the same ratio of carbon-14 to carbon-12 into their tissues and the animals eating the plants have a similar ratio in their tissues. After organisms die, their carbon-14 decays to nitrogen at a certain rate while the carbon-12 content remains constant. Thus, in principle, measuring the ratio of carbon-14 to carbon-12 in the dead organism provides an indication of how long the organism has been dead. This dating method is limited by the 5730 year half-life of carbon-14 to a maximum of 50,000 to 60,000 years. The accuracy of carbon dating has been called into question primarily because the concentration of carbon-14 in the atmosphere varies over time and some plants have the capacity to exclude carbon-14 from their intake.

Radioactive Decay Rates

The decay rate, or activity, of a radioactive substance are characterized by:

1. Constant quantities:

- Half life—symbol $t_{1/2}$ — the time for half of a substance to decay.

- Mean lifetime—symbol τ — the average lifetime of any given particle.

- Decay constant—symbol λ — the inverse of the mean lifetime.

(Note that although these are constants, they are associated with statistically random behavior of substances, and predictions using these constants are less accurate for a small number of atoms.)

2. Time-variable quantities:

- Total activity—symbol A—number of decays an object undergoes per second.

- Number of particles—symbol N—the total number of particles in the sample.

- Specific activity—symbol S_A — number of decays per second per amount of substance. The "*amount of substance*" can be the unit of either mass or volume.

These are related as follows:

$$t_{1/2} = \frac{\ln(2)}{\lambda} = \tau \ln(2)$$

$$A = -\frac{dN}{dt} = \lambda N$$

$$S_A a_0 = -\frac{dN}{dt}\bigg|_{t=0} = \lambda N_0$$

where

a_0 is the initial amount of active substance—substance that has the same percentage of unstable particles as when the substance was formed.

Activity Measurements

The units in which activities are measured are: Becquerel (symbol Bq) = number of disintegrations per second; curie (Ci) = 3.7×10^{10} disintegrations per second. Low activities are also measured in disintegrations per minute (dpm).

Decay Timing

As discussed above, the decay of an unstable nucleus is entirely random and it is impossible to predict when a particular atom will decay. However, it is equally likely to decay at any time. Therefore, given a sample of a particular radioisotope, the number of decay events $-dN$ expected to occur in a small interval of time dt is proportional to the number of atoms present. If N is the number of atoms, then the probability of decay $(- dN/N)$ is proportional to dt:

$$\left(-\frac{dN}{N}\right) = \lambda \cdot dt$$

Particular radionuclides decay at different rates, each having its own decay constant (λ). The negative sign indicates that N decreases with each decay event. The solution to this first-order differential equation is the following function:

$$N(t) = N_0 e^{-\lambda t} = N_0 e^{-t/\tau}$$

This function represents exponential decay. It is only an approximate solution, for two reasons. Firstly, the exponential function is continuous, but the physical quantity N can only take non-negative integer values. Secondly, because it describes a random process, it is only statistically true. However, in most common cases, N is a very large number and the function is a good approximation.

In addition to the decay constant, radioactive decay is sometimes characterized by the mean lifetime. Each atom "lives" for a finite amount of time before it decays, and the mean lifetime is the arithmetic mean of all the atoms' lifetimes. It is represented by the symbol τ, and is related to the decay constant as follows:

$$\tau = \frac{1}{\lambda}$$

A more commonly used parameter is the half-life. Given a sample of a particular radionuclide, the half-life is the time taken for half the radionuclide's atoms to decay. The half life is related to the decay constant as follows:

$$t_{1/2} = \frac{\ln 2}{\lambda} = \tau \ln 2$$

This relationship between the half-life and the decay constant shows that highly radioactive substances are quickly spent, while those that radiate weakly endure longer. Half-lives of known radionuclides vary widely, from more than 10^{19} years (such as for very nearly stable nuclides, for example, ^{209}Bi), to 10^{-23} seconds for highly unstable ones.

Nuclear Binding Energy

Nuclei are made up of protons and neutron, but the mass of a nucleus is always less than the sum of the individual masses of the protons and neutrons which constitute it. The difference is a measure of the nuclear binding energy which holds the nucleus together. This binding energy can be calculated from the Einstein relationship:

$$\text{Nuclear binding energy} = \Delta mc^2$$

For the alpha particle $\Delta m = 0.0304$ u which gives a binding energy of 28.3 MeV.

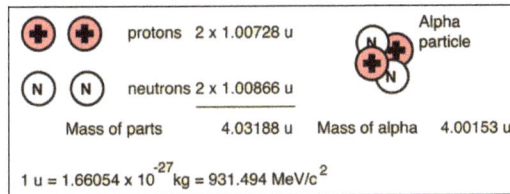

The enormity of the nuclear binding energy can perhaps be better appreciated by comparing it to the binding energy of an electron in an atom. The comparison of the alpha particle binding energy with the binding energy of the electron in a hydrogen atom is shown below. The nuclear binding energies are on the order of a million times greater than the electron binding energies of atoms.

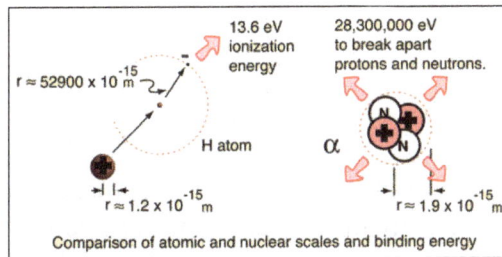

Comparison of atomic and nuclear scales and binding energy

Fission and Fusion can Yield Energy

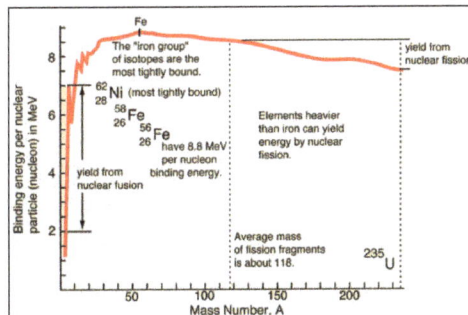

Nuclear Binding Energy Curve

The binding energy curve is obtained by dividing the total nuclear binding energy by the number of nucleons. The fact that there is a peak in the binding energy curve in the region of stability near iron means that either the breakup of heavier nuclei (fission) or the combining of lighter nuclei (fusion) will yield nuclei which are more tightly bound (less mass per nucleon).

The binding energies of nucleons are in the range of millions of electron volts compared to tens of eV for atomic electrons. Whereas an atomic transition might emit a photon in the range of a few electron volts, perhaps in the visible light region, nuclear transitions can emit gamma-rays with quantum energies in the MeV range.

The Iron Limit

The buildup of heavier elements in the nuclear fusion processes in stars is limited to elements below iron, since the fusion of iron would subtract energy rather than provide it. Iron-56 is abundant in stellar processes, and with a binding energy per nucleon of 8.8 MeV, it is the third most tightly bound of the nuclides. Its average binding energy per nucleon is exceeded only by ^{58}Fe and ^{62}Ni, the nickel isotope being the most tightly bound of the nuclides.

Fission and Fusion Yields

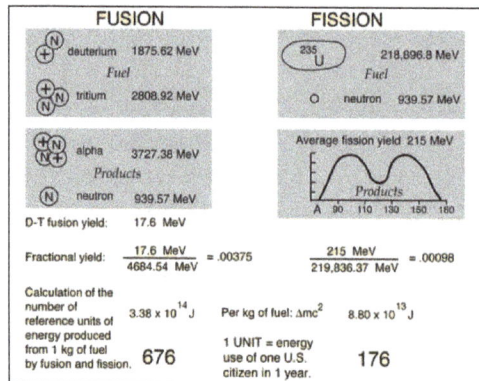

Deuterium-tritium fusion and uranium-235 fission are compared in terms of energy yield. Both the single event energy and the energy per kilogram of fuel are compared. Then they expressed in terms of a nominal per capita U.S. energy use: 5×10^{11} joules. This figure is dated and probably high, but it gives a basis for comparison. The values above are the total energy yield, not the energy delivered to a consumer.

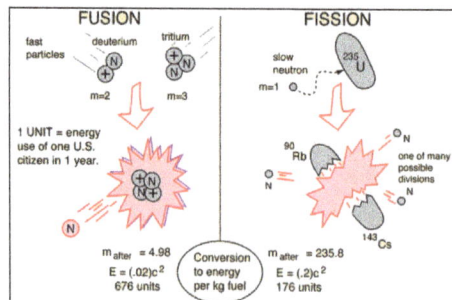

Nuclear Fusion

Nuclear fusion is an atomic reaction in which multiple atom s combine to create a single, more massive atom. The resulting atom has a slightly smaller mass than the sum of the masses of the original atoms. The difference in mass is released in the form of energy during the reaction, according to the Einstein formula $E = mc^2$, where E is the energy in joule s, m is the mass difference in kilogram s, and c is the speed of light (approximately $300,000,000$ or 3×10^8 meters per second).

The most common nuclear fusion reaction in the universe, and the one of most interest to scientists, is the merging of hydrogen nuclei to form helium nuclei. This is the process that occurs in the interiors of stars including the sun. Hydrogen fusion is responsible for the enormous energy output that stars produce. The reaction involves three steps. First, two proton s combine to form a deuterium nucleus, which consists of one proton and one neutron. A positron (also called an anti-electron) and a neutrino (a particle with negligible mass but extreme penetrating power) are generated during this part of the process. Second, the deuterium nucleus combines with another proton, forming a nucleus of helium 3, which consists of two protons and one neutron. An energetic photon is produced during this part of the process, with a wavelength in the gamma-ray portion of the electromagnetic spectrum. Finally, two nuclei of helium 3 combine to form a nucleus of helium 4, which consists of two protons and two neutrons. In this part of the process, two protons (ordinary hydrogen nuclei) are released. These protons can eventually become involved in another fusion reaction.

Nuclear fusion requires extremely high temperatures, on the order of tens of millions of degrees Celsius. In addition, an intense attractive force, such as gravitation of the magnitude that occurs in the centers of stars, is necessary to overcome the electrostatic repulsion among positively charged nuclei. Scientists can generate the high temperatures and forces required to produce uncontrolled hydrogen fusion, the most notable example being the hydrogen bomb. However, sustaining these temperatures and forces indefinitely, in order to construct a hydrogen fusion reactor that can generate useful energy, has proven difficult.

Applications of Nuclear Fusion

We are still at an experimental stage as far as nuclear fusion reactions are concerned.

- Clean: No combustion occurs in nuclear power (fission or fusion), so there is no air pollution.

- Less nuclear waste: The fusion reactors will not produce high-level nuclear wastes like their fission counterparts, so disposal will be less of a problem. In addition, the wastes will not be of weapons-grade nuclear materials as is the case in fission reactors.

If utilized properly, nuclear fusion is the answer to the world's power crisis problem. It is clean and produces a minimal amount of nuclear waste as compared to fission reactions. The fuel for fusion, Deuterium, and Tritium, are also readily available in nature. Scientists are hopeful that in the coming centuries, fusion will be a viable alternative power source.

Nuclear Fission

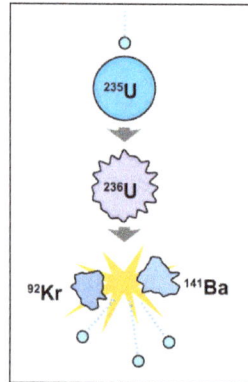

Figure: A model of a fission reaction of uranium-235. Note that
this is just one of the many possible fission reactions.

Nuclear fission is the process of splitting apart nuclei (usually large nuclei). When large nuclei, such as uranium-235, fissions, energy is released. So much energy is released that there is a measurable decrease in mass, from the mass-energy equivalence. This means that some of the mass is converted to energy. The amount of mass lost in the fission process is equal to about 3.20×10^{-11} J of energy. This fission process generally occurs when a large nucleus that is relatively unstable (meaning that there is some level of imbalance in the nucleus between the Coulomb force and the strong nuclear force) is struck by a low energy thermal neutron. In addition to smaller nuclei being created when fission occurs, fission also releases neutrons.

Enrico Fermi originally split the uranium nuclei in 1934. He believed that certain elements could be produced by bombarding uranium with neutrons. Although he expected the new nuclei to have larger atomic numbers than the original uranium, he found that the formed nuclei were radioisotopes of lighter elements. These results were correctly interpreted by Lise Meitner and Otto Frisch over Christmas vacation.

The enormous energy that's released from this splitting comes from how hard the protons are repelling each other with the Coulomb force, barely held together by the strong force. Each proton is pushing every other proton with about 20 N of force, about the force of a hand resting on a person's lap. This is an incredibly huge force for such small particles. This huge force over a small distance leads to a fair amount of released energy which is large enough to cause a measurable reduction in mass. This means that the total mass of each of the fission fragments is less than the mass of the starting nucleus. This missing mass is known as the mass defect.

It is convenient to talk about the amount of energy that binds the nuclei together. All nuclei having this binding energy except hydrogen (which has just 1 proton and no neutrons). It's helpful to think about the binding energy available to each nucleon and this is called the binding energy per nucleon. This is essentially how much energy is required per nucleon to separate a nucleus. The products of fission are more stable, meaning that it is more difficult to split them apart. Since the binding energy per nucleon for fission products is higher, their total nucleonic mass is lower. The result of this higher binding energy and lower mass results in the production of energy. Essentially, mass defect and nuclear binding energy are interchangeable terms.

Use in Energy Generation

Fission of heavier elements is an exothermic reaction. Fission can release up to 200 million eV compared to burning coal which only gives a few eV. From this number alone it is apparent why nuclear fission is used in electricity generation. Additionally, the amount of energy released is much more efficient per mass than that of coal. The main reason that nuclear fission is used for electricity generation is because with proper moderation and the use of control rods, the ejected free neutrons from the fission reaction can then go on to react further with the fuel again. This then creates a sustained nuclear chain reaction, which releases fairly continuous amounts of energy. One downside to the use of fission as a method of generating electricity is the resulting daughter nuclei are radioactive.

When nuclear fission is used to generate electricity, it is referred to as nuclear power. In this case, uranium-235 is used as the nuclear fuel and its fission is triggered by the absorption of a slow moving thermal neutron. Other isotopes that can be induced to fission like this are plutonium-239, uranium-233, and thorium-232. For elements lighter than iron on the periodic table nuclear fusion instead of nuclear fission yields energy. However, currently there is not a method that allows us to access the power that fusion could produce.

Reference

- Nuclear-physics, physics: byjus.com, Retrieved 5 March, 2019

- Radioactive_decay: newworldencyclopedia.org, Retrieved 16 June, 2019

- Nuclear-fusion: techtarget.com, Retrieved 18 January, 2019

- Nuclear-fusion, physics: byjus.com, Retrieved 8 August, 2019

- Nuclear_fission, encyclopedia: energyeducation.ca, Retrieved 7 April, 2019

11

Particle Physics

The branch of physics that studies the nature of the particles that constitute matter and radiation is known as particle physics. Gravitational force, weak force, electromagnetic force, strong forces are the four fundamental forces that fall under the domain of particle physics. The topics elaborated in this chapter will help in gaining a better perspective about particle physics.

Particle physics is a branch of physics that studies the elementary constituents of matter and radiation, and the interactions between them. It is also called high energy physics, because many elementary particles do not occur under normal circumstances in nature, but can be created and detected during energetic collisions of other particles, as is done in particle accelerators.

Our understanding of the nature of physical existence derives in large part from the theories of particle physics. The elementary particles are the ground of existence but there is some mystery attached to how they exist. Described by quantum mechanics, they can be viewed as structureless and dimensionless points or as waves. All other particles are complex entities that derive their three-dimensional existence from the relationships of their constituent elementary particles.

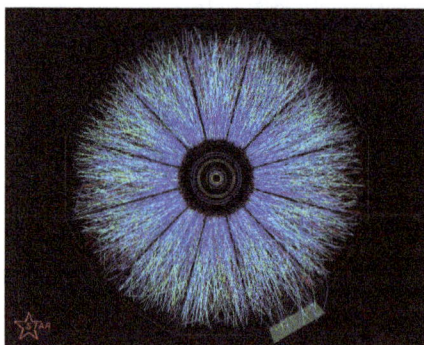

Particles erupt from the collision point of two relativistic (100GeV) gold ions in the STAR detector of the Relativistic Heavy Ion Collider. Electrically charged particles are discernable by the curves they trace in the detector's magnetic field

Subatomic Particles

Modern particle physics research is focused on subatomic particles, which have less structure than atoms. These include matter particles such as the electron, proton, and neutron (protons and neutrons are actually composite particles, made up of quarks), as well as the force-carrying particles, such as photons and gluons and a wide variety of exotic particles.

Strictly speaking, the term particle is something of a misnomer. The objects studied by particle physics obey the principles of quantum mechanics. As such, they exhibit wave-particle duality, displaying particle-like behavior under certain experimental conditions and wave-like behavior in others. Theoretically, they are described neither as waves nor as particles, but as state vectors in an abstract Hilbert space. Following the convention of particle physicists, "elementary particles" is used to refer to objects such as electrons and photons, with the understanding that these "particles" display wave-like properties as well.

All the particles observed to date have been catalogued in a quantum field theory called the Standard Model, which is often regarded as particle physics' best achievement to date. The Standard Model combines quantum electrodynamics and quantum chromodynamics in a coherent conceptual framework that describes the elementary subatomic particles and three of the four interactions. It contains 12 types of matter particles arranged in three generations of increasing energy, and their corresponding antiparticles. The model also describes the strong, weak, and electromagnetic interactions in terms of an exchange of force-carrying particles; gluons, W and Z bosons, and photons. Combinations of the elementary particles account for the hundreds of other species of particles discovered since the 1960s. The Standard Model has been found to agree with almost all the experimental tests conducted to date. However, most particle physicists believe that it is an incomplete description of nature, and that a more fundamental theory awaits discovery. In recent years, measurements of neutrino mass have provided the first experimental deviations from the Standard Model.

Theoretical Particle Physics

Theoretical particle physics attempts to develop the models, theoretical framework, and mathematical tools needed to understand current experiments and make predictions for future experiments. There are several major efforts in theoretical particle physics today and each includes a range of different activities. The efforts in each area are interrelated.

One of the major activities in theoretical particle physics is the attempt to better understand the Standard Model and its tests. By extracting the parameters of the Standard Model from experiments with less uncertainty, this work probes the limits of the Standard Model and therefore expands the understanding of nature. These efforts are made challenging by the difficult nature of calculating many quantities in quantum chromodynamics. Some theorists making these efforts refer to themselves as phenomenologists and may use the tools of quantum field theory and effective field theory. Others make use of lattice field theory and call themselves lattice theorists.

Another major effort is in model building where model builders develop ideas for what physics may lie beyond the Standard Model (at higher energies or smaller distances). This work is often motivated by the hierarchy problem and is constrained by existing experimental data. It may involve work on supersymmetry, alternatives to the Higgs mechanism, extra spatial dimensions (such as the Randall-Sundrum models), Preon theory, combinations of these, or other ideas.

A third major effort in theoretical particle physics is string theory. String theorists attempt to construct a unified description of quantum mechanics and general relativity by building a theory based on small strings and branes rather than particles. If the theory is successful in this, it may be considered a "Theory of Everything."

There are also other areas of work in theoretical particle physics ranging from particle cosmology to loop quantum gravity.

Experimental Laboratories

The world's major particle physics laboratories are:

- Brookhaven National Laboratory (Long Island, United States). Its main facility is the Relativistic Heavy Ion Collider (RHIC), which collides heavy ions such as gold ions and polarized protons. It is the world's first heavy ion collider, and the world's only polarized proton collider.

- Budker Institute of Nuclear Physics (Novosibirsk, Russia). Its main projects are now the electron-positron colliders VEPP-2000, operated since 2006, and VEPP-4, started experiments in 1994. Earlier facilities include the first electron-electron beam-beam collider VEP-1, which conducted experiments from 1964 to 1968; the electron-positron colliders VEPP-2, operated from 1965 to 1974; and, its successor VEPP-2M, performed experiments from 1974 to 2000.

- CERN (European Organization for Nuclear Research) (Franco-Swiss border, near Geneva). Its main project is now the Large Hadron Collider (LHC), which had its first beam circulation on 10 September 2008, and is now the world's most energetic collider of protons. It also became the most energetic collider of heavy ions after it began colliding lead ions. Earlier facilities include the Large Electron–Positron Collider (LEP), which was stopped on 2 November 2000 and then dismantled to give way for LHC; and the Super Proton Synchrotron, which is being reused as a pre-accelerator for the LHC and for fixed-target experiments.

- DESY (Deutsches Elektronen-Synchrotron) (Hamburg, Germany). Its main facility was the Hadron Elektron Ring Anlage (HERA), which collided electrons and positrons with protons. The accelerator complex is now focused on the production of synchrotron radiation with PETRA III, FLASH and the European XFEL.

- Fermi National Accelerator Laboratory (Fermilab) (Batavia, United States). Its main facility until 2011 was the Tevatron, which collided protons and antiprotons and was the highest-energy particle collider on earth until the Large Hadron Collider surpassed it on 29 November 2009.

- Institute of High Energy Physics (IHEP) (Beijing, China). IHEP manages a number of China's major particle physics facilities, including the Beijing Electron–Positron Collider II(BEPC II), the Beijing Spectrometer (BES), the Beijing Synchrotron Radiation Facility (BSRF), the International Cosmic-Ray Observatory at Yangbajing in Tibet, the Daya Bay Reactor Neutrino Experiment, the China Spallation Neutron Source, the Hard X-ray Modulation Telescope (HXMT), and the Accelerator-driven Sub-critical System (ADS) as well as the Jiangmen Underground Neutrino Observatory (JUNO).

- KEK (Tsukuba, Japan). It is the home of a number of experiments such as the K2K experiment, a neutrino oscillation experiment and Belle II, an experiment measuring the CP violation of B mesons.

- SLAC National Accelerator Laboratory (Menlo Park, United States). Its 2-mile-long linear particle accelerator began operating in 1962 and was the basis for numerous electron and positron collision experiments until 2008. Since then the linear accelerator is being used for the Linac Coherent Light Source X-ray laser as well as advanced accelerator design research. SLAC staff continue to participate in developing and building many particle detectors around the world.

Many other particle accelerators also exist. The techniques required for modern experimental particle physics are quite varied and complex, constituting a sub-specialty nearly completely distinct from the theoretical side of the field.

High Energy Physics Compared to Low Energy Physics

The term *high energy physics* requires elaboration. Intuitively, it might seem incorrect to associate "high energy" with the physics of very small, *low* mass objects, like subatomic particles. By comparison, an example of a macroscopic system, one gram of hydrogen, has $\sim 6 \times 10^{23}$ times the mass of a single proton. Even an entire beam of protons circulated in the LHC contains $\sim 3.23 \times 10^{14}$ protons, each with 6.5×10^{12} eV of energy, for a total beam energy of $\sim 2.1 \times 10^{27}$ eV or ~ 336.4 MJ, which is still $\sim 2.7 \times 10^{5}$ times lower than the mass-energy of a single gram of hydrogen. Yet, the macroscopic realm is "low energy physics", while that of quantum particles is "high energy physics".

The interactions studied in other fields of physics and science have comparatively very low energy. For example, the photon energy of visible light is about 1.8 to 3.1 eV. Similarly, the bond-dissociation energy of a carbon–carbon bond is about 3.6 eV. Other chemical reactions typically involve similar amounts of energy. Even photons with far higher energy, gamma rays of the kind produced in radioactive decay, mostly have photon energy between 10^5 eV and 10^7 eV – still two orders of magnitude lower than the mass of a single proton. Radioactive decay gamma rays are considered as part of nuclear physics, rather than high energy physics.

The proton has a mass of around 9.4×10^{8} eV; some other massive quantum particles, both elementary and hadronic, have yet higher masses. Due to these very high energies *at the single particle level*, particle physics is, in fact, high-energy physics.

The Future of Particle Physics

Particle physicists internationally agree on the most important goals of particle physics research in the near and intermediate future. The overarching goal, which is pursued in several distinct ways, is to find and understand what physics may lie beyond the Standard Model. There are several powerful experimental reasons to expect new physics, including dark matter and neutrino mass. There are also theoretical hints that this new physics should be found at accessible energy scales. Most importantly, though, there may be unexpected and unpredicted surprises that will give the most opportunity to learn about nature.

Much of the efforts to find this new physics are focused on new collider experiments. A (relatively) near-term goal is the completion of the Large Hadron Collider (LHC) in 2007, which will continue the search for the Higgs boson, supersymmetric particles, and other new physics. An intermediate goal is the construction of the International Linear Collider (ILC), which will complement the LHC by allowing more precise measurements of the properties of newly found particles. A decision for the technology of the ILC has been taken in August 2004, but the site has still to be agreed upon.

Additionally, there are important non-collider experiments that also attempt to find and understand physics beyond the Standard Model. One important non-collider effort is the determination of the neutrino masses since these masses may arise from neutrinos mixing with very heavy particles. In addition, cosmological observations provide many useful constraints on the dark matter, although it may be impossible to determine the exact nature of the dark matter without the colliders. Finally, lower bounds on the very long life time of the proton put constraints on Grand Unification Theories at energy scales much higher than collider experiments will be able to probe anytime soon.

Four Fundamental Forces

The four fundamental forces of nature are gravitational force, weak nuclear force, electromagnetic force and strong nuclear force. The weak and strong forces are effective only over a very short range and dominate only at the level of subatomic particles. Gravity and electromagnetic force have infinite range.

The Four Fundamental Forces and their Strengths

1. Gravitational Force – Weakest force; but infinite range. (Not part of standard model)

2. Weak Nuclear Force – Next weakest; but short range.

3. Electromagnetic Force – Stronger, with infinite range.

4. Strong Nuclear Force – Strongest; but short range.

Gravitational Force

The gravitational force is weak, but very long ranged. Furthermore, it is always attractive. It acts between any two pieces of matter in the Universe since mass is its source.

Weak Nuclear Force

The weak force is responsible for radioactive decay and neutrino interactions. It has a very short range and. As its name indicates, it is very weak. The weak force causes Beta decay ie. the conversion of a neutron into a proton, an electron and an antineutrino.

Electromagnetic Force

The electromagnetic force causes electric and magnetic effects such as the repulsion between like electrical charges or the interaction of bar magnets. It is long-ranged, but much weaker than the strong force. It can be attractive or repulsive, and acts only between pieces of matter carrying electrical charge. Electricity, magnetism, and light are all produced by this force.

Strong Nuclear Force

The strong interaction is very strong, but very short-ranged. It is responsible for holding the nuclei of atoms together. It is basically attractive, but can be effectively repulsive in some circumstances.

The strong force is 'carried' by particles called gluons; that is, when two particles interact through the strong force, they do so by exchanging gluons. Thus, the quarks inside of the protons and neutrons are bound together by the exchange of the strong nuclear force.

While they are close together the quarks experience little force, but as they separate the force between them grows rapidly, pulling them back together. To separate two quarks completely would require far more energy than any possible particle accelerator could provide.

Fundamental Force Particles

Force	Particles Experiencing	Force Carrier Particle	Range	Relative Strength*
Gravity acts between objects with mass	all particles with mass	graviton (not yet observed)	infinity	much weaker
Weak Force governs particle decay	quarks and leptons	W^+, W^-, Z^0 (W and Z)	short range	
Electromagnetism acts between electrically charged particles	electrically charged	γ (photon)	infinity	
Strong Force** binds quarks together	quarks and gluons	g (gluon)	short range	much stronger

Electroweak Theory and Grand Unification Theories (GUT)

There is a speculation, that In the very early Universe when temperatures were very high (the Planck Scale) all four forces were unified into a single force. Then, as the temperature dropped, gravitation separated first and then the other 3 forces separated. Even then, the weak, electromagnetic, and strong forces were unified into a single force. When the temperature dropped these forces got separated from each other, with the strong force separating first and then at a still lower temperature the electromagnetic and weak forces separating to leave us with the 4 distinct forces that we see in our present Universe. The process of the forces separating from each other is called spontaneous symmetry breaking.

- The weak and electromagnetic interactions have been unified under Standard Electroweak Theory, or sometimes just the Standard Model. (Glashow, Weinberg, and Salaam were awarded the Nobel Prize for this in 1979).

- Grand unification theories attempt to treat both strong and electroweak interactions under the same mathematical structure. [Unification of Weak forces and strong forces] PS: Attempts to include gravitation in this picture have not yet been successful.

- Theories that add gravity to the mix and try to unify all four fundamental forces into a single force are called Superunified Theories.

- PS: Grand Unified and Superunified Theories remain theoretical speculations that are as yet unproven, but there is strong experimental evidence for the unification of the electromagnetic and weak interactions in the Standard Electroweak Theory. Furthermore, although GUTs are not proven experimentally, there is strong circumstantial evidence to suggest that a theory at least like a Grand Unified Theory is required to make sense of the Universe.

Elementary Particles

Elementary particles are the smallest known building blocks of the universe. They are thought to have no internal structure, meaning that researchers think about them as zero-dimensional points that take up no space. Electrons are probably the most familiar elementary particles, but the Standard Model of physics, which describes the interactions of particles and almost all forces, recognizes 10 total elementary particles.

A given particle may not necessarily be subject to all four interactions. Neutrinos, for example, experience only the weak and gravitational interaction.

The fundamental particles may be classified into groups in several ways. First, all particles are classified into fermions, which obey Fermi-Dirac statistics and bosons, which obey Bose-Einstein statistics. Fermions have half-integer spin, while bosons have integer spin. All the fundamental fermions have spin 1/2. Electrons and nucleons are fermions with spin 1/2. The fundamental bosons have mostly spin 1. This includes the photon. The pion has spin 0, while the graviton has spin 2. There are also three particles, the W^+, W^- and Z_0 bosons, which are spin 1. They are the carriers of the weak interactions.

We can also classify the particles according to their interactions.

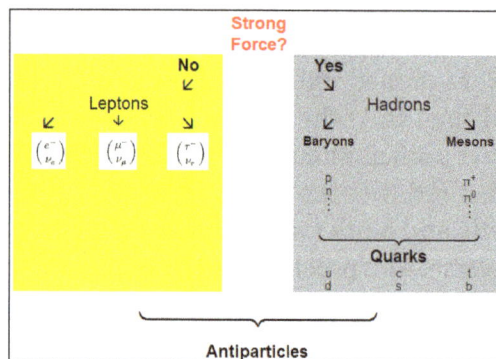

The electron and the neutrino are members of a family of leptons. Originally leptons meant "light particles", as opposed to baryons, or heavy particles, which referred initially to the proton and neutron. The pion, or pi-meson, and another particle called the muon or mu-meson, were called mesons, or medium-weight particles, because their masses, a few hundred times heavier than the electron but six times lighter than a proton, were in the middle. But that distinction turned out not to be very useful. We now recognize the muon to be almost the same as an electron, and the leptons now consist of three "generations" of pairs of particles,

$$\begin{pmatrix} e^- \\ v_e \end{pmatrix}, \begin{pmatrix} \mu^- \\ v_\mu \end{pmatrix}, \begin{pmatrix} \tau^- \\ v_r \end{pmatrix},$$

With the heaviest of these, the tau lepton τ^-, being almost twice as massive as the proton.

The leptons are distinguished from other particles called hadrons in that leptons do not participate in strong interactions. The bottom lepton in each of the three "doublets" shown above is not only neutral, but also has a very small mass. Neutrinos had been considered massless for many years, but more recent experiments have shown their mass to be non-zero.

Table of Leptons

particle			associated neutrino		
Name	Charge (e)	Mass(MeV)	Name	Charge (e)	Mass (MeV)
Electron $\left(e^-\right)$	-1	0.511	Electron neutrino $\left(\nu_e\right)$	0	< 0.000003
Muon $\left(\mu^-\right)$	-1	105.6	Muon neutrino $\left(\nu_\mu\right)$	0	< 0.19
Tau $\left(\tau^-\right)$	-1	1777	Tau neutrino $\left(\nu_\tau\right)$	0	< 18.2

Hadrons are strongly interacting particles. They are divided into baryons and mesons. The baryons are a class of fermions, including the proton and neutron, and other particles which in a decay always produce another baryon, and ultimately a proton. The mesons, are bosons. In addition to the pion, there are other spin 0 particles, four kaons and two eta mesons, and a number of spin one hadrons, including the three rho mesons, which like the pion come in charges 1 and 0. Mesons can decay without necessarily producing other hadrons.

Table of some baryons

Particle	Symbol	Quark Content	Mass MeV/c²	Mean lifetime (s)	Decays to
Proton	p	uud	938.3	Stable	Unobserved
Neutron	n	ddu	939.6	885.7 ± 0.8	$p + e^- + \bar{\nu}_e$
Delta	Δ^{++}	uuu	1232	6×10^{-24}	$\pi^+ + p$
Delta	Δ^+	uud	1232	6×10^{-24}	$\pi^+ + n$ or $\pi^0 + p$
Delta	Δ^0	udd	1232	6×10^{-24}	$\pi^0 + n$ or $\pi^- + p$
Delta	Δ^-	ddd	1232	6×10^{-24}	$\pi^- + n$
Lambda	Λ^0	uds	1115.7	2.60×10^{-10}	$\pi^- + p$ or $\pi^0 + n$
Sigma	Σ^+	uus	1189.4	0.8×10^{-10}	$\pi^0 + p$ or $\pi^+ + n$
Sigma	Σ^0	uds	1192.5	6×10^{-20}	$\Lambda^0 + \gamma$
Sigma	Σ^-	dds	1197.4	1.5×10^{-10}	n
Xi	Ξ^0	uss	1315	2.9×10^{-10}	$\Lambda^0 + \pi^0$
Xi	Ξ^-	dss	1321	1.6×10^{-10}	$\Lambda^0 + \pi^-$
Omega	Ω^-	sss	1672	0.82×10^{-10}	$\Lambda^0 + K^-$ or $\Xi^0 + \pi^-$

Table of some mesons

Particle	Symbol	Anti-particle	Quark Content	Mass MeV/c²	Mean lifetime (s)	Principal decays
Charged Pion	π^+	π^-	$u\bar{d}$	139.6	2.60×10^{-8}	$\mu^+ + \nu_\mu$
Neutral Pion	π^0	Self	$u\bar{u}-d\bar{d}$	135.0	0.84×10^{-16}	2γ
Charged Kaon	K^+	K^-	$u\bar{s}$	493.7	1.24×10^{-8}	$\mu^+ + \nu_\mu$ or $\pi^+ + \pi_0$
Neutral Kaon	K^0	K^0	$d\bar{s}$	497.7		
Eta	η	Self	$uu+dd-2ss$	547.8	5×10^{-19}	
Eta Prime	η'	Self	$uu+dd+ss$	957.6	3×10^{-21}	

Each elementary particle is associated with an antiparticle with the same mass and opposite charge. Some particles, such as the photon, are identical to their antiparticle. Such particles must be neutral, but not all neutral particles are identical to their antiparticle. Particle-antiparticle pairs can annihilate each other if they are in appropriate quantum states, releasing an amount of energy equal to twice the rest energy of the particle. They can also be produced in various processes, if enough energy is available. The minimum amount of energy needed is twice the rest energy of the particle, if momentum conservation allows the particle-antiparticle pair to be produced at rest. Most often the antiparticle is denoted by the same symbol as the particle, but with a line over the symbol. For example, the antiparticle of the proton p, is denoted by $\bar{p}$.

Protons and neutrons are made of still smaller particles called quarks. At this time it appears that the two basic constituents of matter are the leptons and the quarks. There are believed to be six types of each. Each quark type is called a flavor, there are six quark flavors. Each type of lepton and quark also has a corresponding antiparticle, a particle that has the same mass but opposite electrical charge and magnetic moment. An isolated quark has never been found, quarks appear to almost always be found in pairs or triplets with other quarks and antiquarks. The resulting particles are the hadrons, more than 200 of which have been identified. Baryons are made up of 3 quarks, and mesons are made up of a quark and an anti-quark. Baryons are fermions and mesons are bosons. Two theoretically predicted five-quark particles, called pentaquarks, have been produced in the laboratory. Four- and six-quark particles are also predicted but have not been found.

The six quarks have been named up, down, charm, strange, top, and bottom. The top quark, which has a mass greater than an entire atom of gold, is about 35 times more massive than the next biggest quark and may be the heaviest particle nature has ever created. The quarks found in ordinary matter are the up and down quarks, from which protons and neutrons are made. A proton consists of two up quarks and a down quark, and a neutron consists of two down quarks and an up quark. The pentaquark consists of two up quarks, two down quarks, and the strange antiquark. Quarks have fractional charges of one third or two thirds of the basic charge of the electron or proton. Particles made from quarks always have integer charge.

Table of Quarks

Name	Symbol	Charge (e)	Spin	Mass MeV/c²	Strangeness	Baryon number	Lepton number
up	u	+2/3	1/2	1.7-3.3	0	1/3	0
down	d	-1/3	1/2	4.1-5.8	0	1/3	0
strange	s	-1/3	1/2	101	-1	1/3	0
charm	c	+2/3	1/2	1270	0	1/3	0
bottom	b	-1/3	1/2	4190-4670	0	1/3	0
top	t	+2/3	1/2	172000	0	1/3	0

In the current theory, known as the Standard Model there are 12 fundamental matter particle types and their corresponding antiparticles. In addition, there are gluons, photons, and W and Z bosons, the force carrier particles that are responsible for strong, electromagnetic, and weak interactions respectively. These force carriers are also fundamental particles.

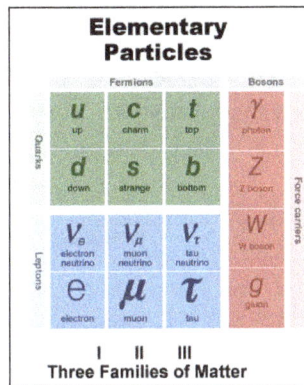

All we know is that quarks and leptons are smaller than 10-19 meters in radius. As far as we can tell, they have no internal structure or even any size. It is possible that future evidence will, once again, show our understanding to be incomplete and demonstrate that there is substructure within the particles that we now view as fundamental.

The Discovery of Elementary Particles

The first subatomic particle to be discovered was the electron, identified in 1897 by J. J. Thomson. After the nucleus of the atom was discovered in 1911 by Ernest Rutherford, the nucleus of ordinary hydrogen was recognized to be a single proton. In 1932 the neutron was discovered. An atom was seen to consist of a central nucleus containing protons and neutrons, surrounded by orbiting electrons. However, other elementary particles not found in ordinary atoms immediately began to appear.

In 1928 the relativistic quantum theory of P. A. M. Dirac predicted the existence of a positively charged electron, or positron, which is the antiparticle of the electron. It was first detected in 1932. Difficulties in explaining beta decay led to the prediction of the neutrino in 1930, and by 1934 the existence of the neutrino was firmly established in theory, although it was not actually detected until 1956. Another particle was also added to the list, the photon, which had been first suggested by Einstein in 1905 as part of his quantum theory of the photoelectric effect.

The next particles discovered were related to attempts to explain the strong interactions, or strong nuclear force, binding nucleons together in an atomic nucleus. In 1935 Hideki Yukawa suggested that a meson, a charged particle with a mass intermediate between those of the electron and the proton, might be exchanged between nucleons. The meson emitted by one nucleon would be absorbed by another nucleon. This would produce a strong force between the nucleons, analogous to the force produced by the exchange of photons between charged particles interacting through the electromagnetic force. It is now known that the strong force is mediated by the gluon. The following year a particle of approximately the required mass, about 200 times that of the electron, was discovered and named the mu-meson, or muon. However, its behavior did not conform to that of the theoretical particle. In 1947 the particle predicted by Yukawa was finally discovered and named the pi-meson, or pion. Both the muon and the pion were first observed in cosmic rays. Further studies of cosmic rays turned up more particles. By the 1950s these elementary particles were also being observed in the laboratory as a result of particle collisions in particle accelerators.

By the early 1960s over 30 "fundamental particles" had been found. A rigorous way of classifying them was needed. Were there any symmetries or patterns? Murray Gell-Mann believed that a framework for such patterns could be found in the mathematical structure of groups. A symmetry group called SU(3) offered patterns he was looking for. In 1961, after grouping the known particles, he predicted the existence of the η particle which was needed to complete a pattern. The η particle was discovered a few months later.

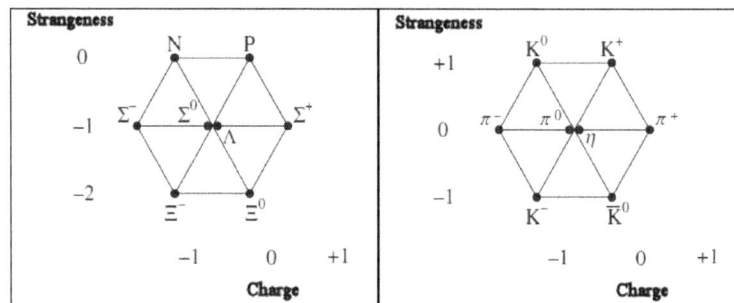

Example patterns for some baryons and mesons (The Eightfold Way)

After finding the patterns an explanation was needed. The patterns could be produced if the known particles were viewed as combinations of 3 fundamental subunits with fractional charge, the up, down, and strange quarks and their antiquarks. There were however problems with the Pauli Exclusion Principle. The quarks are spin 1/2 fermions and the Δ^{++} (uuu) and the Ω^{-} (sss) seemed to contain at least two quarks with exactly the same quantum numbers. The quark theory was not really accepted until deep inelastic scattering experiments revealed structure inside the protons in the later 1960s.

The charm quark was discovered in 1974, the bottom quark in 1977, and the top quark in 1995. The tau particle was detected in a series of experiments between 1974 and 1977 and the discovery of the tau neutrino was announced in 2000. It was the last of the particles in the Standard Model of elementary particles to be detected.

One of the current frontiers in the study of elementary particles concerns the interface between that discipline and cosmology. The known quarks and leptons, for instance, are typically grouped in three families, where each family contains two quarks and two leptons. Investigators have wondered whether additional families of elementary particles might be found. Recent work in cosmology pertaining to the evolution of the universe has suggested that there could be no more families than four, and the cosmological theory has been substantiated by experimental work at the Stanford Linear Accelerator and at CERN, which indicates that there are no families of elementary particles other than the three that are known today. For example, detailed studies of Z_0 decays at CERN revealed that there can be no more than three different kinds of neutrino. If there was a fourth, or a fifth, further decay routes would be open to the Z_0 which would affect its measured lifetime.

Conservation Laws

Leptons carry an additive quantum number called the lepton number L. The leptons listed in the table above carry L = 1 and their antiparticles carry L = -1. The electron number carried by electrons and electron neutrinos seems to an additive quantum number which is conserved in interactions. Electrons and electron neutrinos have electron number 1 and positrons and electron anti-neutrinos carry electron number -1. Muons, or μ-mesons, behave similarly as electrons. They only have electromagnetic and weak interactions. Only their mass, 106 MeV/c², distinguishes them from electrons. Along with v_μ they carry an additive quantum number, themuon number, that which also seems to be conserved in interactions. Muons decay as

$$\mu^+ --> e^+ + v_\mu + v_e$$

in about 10^{-6}s. The tau and its neutrino carry an additive quantum number as well, which seems to be conserved in interactions. We say that the lepton family number LF also seems to be conserved. However if neutrinos have mass and can change flavor, for example, if muon neutrinos can change into electron neutrinos and vice versa, then only L is conserved.

Are there more such additive quantum numbers? Yes, there is a group of particles called baryons, and a corresponding conserved quantum number called baryon number B. Baryons have baryon number B = 1 and anti-baryons have baryon number B = -1. The lightest baryon is the proton, and it is the only stable baryon. Since the neutron decays by $n --> p + e^- + v_e$ and the electron and anti-neutrino are leptons, not baryons, Bconservation requires that the neutron is also baryon.

It is fairly easy to spot a baryon in a table of elementary particles. Suppose you are looking at a particle which might be a baryon. If it is not the proton and it is a baryon, it must decay. Baryon conservation then requires a baryon among the decay products, although you may not know which of the decay products is the baryon. Let all of the decay products themselves decay. The baryon's decay yields another baryon. Keep going until all the particles are stable. Among all the resulting particles there must be one net baryon. Since the proton is the only stable baryon, that baryon must be a proton. Hence, a particle is a baryon if and only if there is one net proton among its ultimate decay products.

Baryons are made up of 3 quarks. All quarks have baryon number B = 1/3, and all anti-quarks have baryon number B = -1/3.

Everything else in the table besides the baryons and leptons is called a meson. Mesons are made up of a quark and an anti-quark. Mesons have L = 0 and B = 0, and they have no net leptons or baryons in their ultimate decay products. The number of mesons is not conserved, so there is no "meson number."

After the discovery of pions other particles were discovered in rapid pace. K-mesons (m = 494 MeV/c²) are similar to π -mesons, they do not carry a nucleon number. The K-meson decays in the following way.

$$K^+ ---> \pi^+ + \pi^0$$

Hyperons are different. The lightest hyperon is the Λ^0 with m = 1116 MeV/c². It decays in the following way

$$\Lambda^0 ---> p + \pi^-, \text{ or } n + \pi^0$$

A hyperons is a baryon.

As better accelerators became available the reaction below was observed

$$\pi^+ + n ---> \Lambda^0 + K^+$$

Obviously in the above process baryon number is conserved. It was remarkable that reactions like

$$\pi^+ + n ---> \Lambda^0 + \pi^+$$

were never observed. This prompted the introduction of a new quantum number, strangeness, by Gell-Mann and Pais. The strangeness $S(\Lambda^0) = -1$, $S(K^+) = 1$. Protons, neutron an pions have no strangeness. In decay processes involving the strong interaction strangeness is conserved. In decay processes involving the weak interaction, such as K-decay or hyperon decay, strangeness is not conserved. This was the first case that some quantum numbers were conserved in strong interactions and electromagnetic interactions, but not in weak interactions. Decay processes governed by the strong interaction can be distinguished easily from decay processes governed by the weak interaction. Characteristic reaction times for the former are on the order of 10^{-23} s while for the latter they are on the order of 10^{-10} s .

Reference

- Particle_physics: newworldencyclopedia.org, Retrieved 2 June, 2019

- Four-fundamental-forces-of-nature: clearias.com, Retrieved 23 February, 2019

- Fundamental-elementary-particles: livescience.com, Retrieved 27 August, 2019

- Particle_classification: phys.utk.edu, Retrieved 17 May, 2019

Permissions

Index

www.ingramcontent.com/pod-product-compliance
Lightning Source LLC
Chambersburg PA
CBHW061305190326

41458CB00011B/3770